Biomarkers in Clinical Drug Development

DRUGS AND THE PHARMACEUTICAL SCIENCES

Executive Editor

James Swarbrick
PharmaceuTech, Inc.
Pinehurst, North Carolina

DRUGS AND THE PHARMACEUTICAL SCIENCES

A Series of Textbooks and Monographs

Biomarkers in Clinical Drug Development

edited by

John C. Bloom
Lilly Research Laboratories
Indianapolis, Indiana, U.S.A.

Robert A. Dean
Lilly Research Laboratories
and Indiana University School of Medicine
Indianapolis, Indiana, U.S.A.

MARCEL DEKKER, INC.   NEW YORK · BASEL

Library of Congress Cataloging-in-Publication Data
A catalog record for this book is available from the Library of Congress.

ISBN: 0-8247-4206-2

This book is printed on acid-free paper.

Headquarters
Marcel Dekker, Inc.
270 Madison Avenue, New York, NY 10016
Tel: 212-696-9000; fax: 212-685-4540

Eastern Hemisphere Distribution
Marcel Dekker AG
Hutgasse 4, Postfach 812, CH-4001 Basel, Switzerland
tel: 41-61-260-6300; fax: 41-61-260-6333

World Wide Web
http://www.dekker.com

The publisher offers discounts on this book when ordered in bulk quantities. For more information, write to Special Sales/Professional Marketing at the headquarters address above.

Current printing (last digit):
10 9 8 7 6 5 4 3 2 1

PRINTED IN THE UNITED STATES OF AMERICA

Preface

The term *biomarker* has the distinction of having become, within a few years of usage, both a euphemism so broadly used so as to limit its usefulness and a holy grail for drug hunters. A biomarker is, in the most general sense, an indicator of a normal or abnormal biological process. A *clinical biomarker* is a relatively new term for such indicators in humans, as applied to healthcare or disease management for individuals or populations, It is a convenient label for novel, or experimental, esoteric and established tests, or clinical assessments that range from a DNA sequence to a self-administered blood pressure measurement to the color of one's eyes.

Clinical biomarkers have always been important in drug development and are central to the demonstration of safety and efficacy. Their inclusion as a requirement on most lists of critical success factors for candidate drugs in development and as a key driver for investment strategies within large research and development organizations stems from the recognition of one simple fact: the ability to demonstrate quickly and precisely the safety and efficacy of new molecular entities early in development will, more than any other factor, define the probability of technical success and the overall value of the pipeline. This can best be accomplished through the strategic application of appropriate biomarkers. Biomarkers are among the fundamental tools applied across all phases of drug development. The data generated by their assessment accounts for up to 80% of those included in New Drug Applications. Despite this sobering fact, relatively little has been written on the application of biomarkers to drug development in textbooks, scientific/trade reviews, and guidance documents from regulatory agencies. While much is provided in the scientific literature on the use of novel and routine biomarkers for health maintenance and disease intervention, missing are informed perspectives and advice on the unique challenges drug hunters face in the application of these increasingly sophisticated tools to the development, registration, and commercialization of pharmaceuticals in today's environment. These challenges relate to the highly regulated environment in which this clinical research is conducted; the use of markers that must simultaneously serve both patient and clinical trial management needs and objectives; the need for predictive value and validation standards that exceed those of markers applied to routine health care delivery (i.e., the impact of false positives and false negatives); and the overarching need to demonstrate and confirm that pharmacokinetic, pharmacodynamic, and safety standards are met for sophisticated candidate interventions that are accessible to a large, electric segment of our global society that chooses to have these agents in their medicine cabinets.

Biomarkers in Clinical Drug Development is an attempt to fill this void. The book approaches the subject from three perspectives: (1) biomarker applied science, (2) strategic applications, and (3) clinical operations. As is the case with most texts on applied science, there has been considerable progress in this field between the inception of this work, its execution, and publication. Most notable have been advances in biomarker discovery, including pharmacogenomics, proteomic applications, and molecular imaging; and regulatory guidance in areas such as cardiac and hepatic safety. Our distinguished contributors have tried to highlight the principles, challenges, and technical approaches that will endure in an environment of rapid change.

We would like to acknowledge the financial, intellectual, and moral support that Eli Lilly's research and development community and management provided, as well as the efforts of our assistant, Ms. Vickie Cafaro, without which this book would not have been possible.

John C. Bloom
Robert A. Dean

Contents

Contributors

Angela Berns Amgen Inc., Thousand Oaks, California, U.S.A.

Luc Bijnens Janssen Research Foundation, Beerse, Belgium

John C. Bloom Diagnostic and Experimental Medicine, Lilly Research Laboratories, Indianapolis, Indiana, U.S.A.

Ronald R. Bowsher Drug Disposition, Lilly Research Laboratories, Indianapolis, Indiana, U.S.A.

Tomasz Burzykowski Limburts Universitair Centrum, Diepenbeek, Belgium

Marc Buyse International Drug Development Institute, Cambridge, Massachusetts, U.S.A.

Wayne A. Colburn MDS Pharma Services, Phoenix, Arizona, U.S.A.

Robert A. Dean Diagnostic and Experimental Medicine and Clinical Diagnostic Services, Lilly Research Laboratories and Departments of Pathology and Laboratory Medicine and Biochemistry and Molecular Biology, Indiana University School of Medicine, Indianapolis, Indiana, U.S.A.

Gregory J. Downing Office of Science Policy and Planning, National Institutes of Health, Bethesda, Maryland, U.S.A.

Helena Geys Limburgs Universitair Centrum, Diepenbeek, Belgium

William J. Groh Indiana University School of Medicine, Indianapolis, Indiana, U.S.A.

Jacqueline Hevy Amgen Inc., Thousand Oaks, California, U.S.A.

Richard D. Hockett   Eli Lilly and Company, Indianapolis, Indiana, U.S.A.

Gordon F. Kapke   Worldwide Technical Affairs, Covance Central Laboratory Services, Indianapolis, Indiana, U.S.A.

Sandra C. Kirkwood   Eli Lilly and Company, Indianapolis, Indiana, U.S.A.

Linda F. Knoob   Eli Lilly and Company, Indianapolis, Indiana, U.S.A.

Jean W. Lee   Bioanalytical Development, MDS Pharma Services, Lincoln, Nebraska, U.S.A.

Barbara Maley   Eli Lilly and Company, Indianapolis, Indiana, U.S.A.

Geert Molenberghs   Limburgs Universitair Centrum, Diepenbeek, Belgium

Gerald D. Nordblom   Pfizer Global Research and Development, Ann Arbor, Michigan, U.S.A.

Charles G. Peterfy   Synarc, Inc., San Francisco, California, U.S.A.

Didier Renard   Limburgs Universitair Centrum, Diepenbeek, Belgium

Gregory D. Sides   Cardiovascular Medical, Lilly Research Laboratories, Eli Lilly and Company, Indianapolis, Indiana, U.S.A.

Frank D. Sistare   Division of Applied Pharmacology Research, Center for Drug Evaluation and Research, Food and Drug Administration, Laurel, Maryland, U.S.A.

Wendell C. Smith   Statistical/Math Sciences, Lilly Research Laboratories, Greenfield, Indiana, U.S.A.

Tony Vangeneugden   Janssen Research Foundation, Beerse, Belgium

1

Biomarkers in Clinical Drug Development: Definitions and Disciplines

John C. Bloom
Lilly Research Laboratories, Indianapolis, Indiana, U.S.A.

I. INTRODUCTION

Recent developments in biomedical science have enabled our health care professionals to diagnose, characterize, and predict disease in ways no one would have thought possible just 10 years ago. The rapid translation of discoveries in the biology and chemistry research laboratories to patient care has brought a dizzying array of breakthrough technologies to the bedside. The new molecular diagnostic, imaging, flow cytometric, and many other tools now used to characterize events such as receptor binding, gene activation, enzyme activity, and organ function are being combined with the better-established diagnostic tests to rapidly define the susceptibility, status, and appropriate interventions of disease in both individual patients and populations. This confluence of novel technologies with the conventional diagnostic armamentarium has required broader nomenclature. The term "biological markers," or "biomarkers," quickly served this purpose and became part of the new jargon of experimental medicine [1,2]. This terminology and the thinking behind it has been particularly useful in the disciplines of therapeutic research, particularly pharmaceutical research and development, where powerful discovery and screening technologies such as high-throughput screening, combinatorial chemistry, tandam mass spectrometry, and DNA microarrays have yielded an avalanche of new molecular entities requiring sophisticated

Table 1 Examples of Biomarkers Used in Clinical Drug Development

Routine clinical pathology	Magnetic resonance imaging (MRI)
Histopathology	Positron emission tomography (PET)
Special laboratory biomarkers	Ultrasound (quantitative, transvaginal, etc.)
Electrocardiogram (ECG)	Angiography
Holter monitoring	Echocardiography
Computed Tomography (CT)	Thallium scans
Bone mineral density (BMD)	Mammography
Routine radiographs (x-rays)	Total body composition

technologies, strategies, and *biomarkers* to demonstrate target activity, safety, and efficacy.

Biomarkers have always been important in clinical development and provide the most practical means of demonstrating that a candidate drug is safe and effective in a disease target population. Clinical laboratory measurements alone now represent 50–80% of the data in a New Drug Application (NDA). Examples of biomarkers commonly used in clinical drug development are listed in Table 1.

II. DEFINITIONS AND APPLICATIONS

A working definition of biological markers, or biomarkers, as applied to drug development was recently proposed by a Biomarker Definitions Working Group (BDWG) [2]. They define a biomarker as "a characteristic that is objectively measured and evaluated as an indicator of normal biological processes, pathogenic processes or pharmacologic responses to a therapeutic intervention." Applications to drug development, defined by the Working Group, include use in early-phase clinical trials to establish "proof of concept"; as diagnostic tools for identifying patients with a specific disease; as tools for characterizing or staging disease processes; as an indicator of disease progress; and for predicting and monitoring the clinical response to therapeutic intervention.

Depending on the application, other terms have been applied to biomarkers, which have been used to classify them [3]. *Natural history markers* are those that measure disease predisposition, severity, or outcome. These are often used to define inclusion or exclusion criteria for patients considered for enrollment into clinical trials, or for stratifying these patient populations. Examples include the genotypes for Factor V Leiden and Apo E as risk factors for thromboembolism and Alzheimer's disease, respectively, and the peptide α-fetoprotein for progression of certain neoplasias.

Markers that reflect a response to therapy or drug treatment are called *drug activity markers*. They are used to demonstrate proof of concept, to establish dose regimens, and for optimizing combination therapies. They are used to measure the pharmacodynamic response, where the magnitude of the change defines the potency of the candidate drug. Examples include biochemical markers for bone resorption and deposition in studies on osteoporosis, such as osteocalcin, bone-specific alkaline phosphatase, and type I collagen propeptides and telopeptides [4]; and viral load (hepatitis or human immunodeficiency virus) for measuring response to antiviral therapy [3].

Finally, biomarkers represent *clinical endpoints* when they measure directly how a patient feels or functions. They may also measure survival. In those rare circumstances where a marker may actually be substituted for a clinical endpoint, they are called *surrogate endpoints*. They are used to predict clinical benefit, or safety, based on epidemiological, therapeutic, or pathophysiological evidence. The use of *blood pressure* (hypertension) as a surrogate for stroke, or measure of risk for that event, has become a classic and rare example of a true surrogate endpoint [5].

Surrogate endpoints in clinical trials are often applied later in clinical practice to monitor disease progression or response to treatment. Use of the term "surrogate endpoint" requires that the clinical endpoints substituted by this marker will be well defined, along with the class of therapeutic intervention being applied and characteristics of the target disease and patient population involved [2]. The BDWG discourages use of the term "surrogate marker," as surrogate literally means "to substitute," and surrogate marker therefore suggests substitution for a marker rather than for a clinical endpoint. The use of surrogate endpoints to demonstrate efficacy of a candidate drug in registration trials, and thereby accelerate marketing approval, has been codified by the U.S. Food and Drug Administration (FDA) in the following regulation [6]:

> *Approval based on a surrogate endpoint or on an effect on a clinical endpoint other than survival or irreversible morbidity.* FDA may grant marketing approval for a new drug product on the basis of adequate and well-controlled clinical trials establishing that the drug product has an effect on a surrogate endpoint that is reasonably likely, based on epidemiologic, therapeutic, pathophysiologic, or other evidence, to predict clinical benefit or on the basis of an effect on a clinical endpoint other than survival or irreversible morbidity. Approval under this section will be subject to the requirement that the applicant study the drug further, to verify and describe its clinical benefit where there is uncertainty as to the relation of the surrogate endpoint to clinical benefit, or of the observed clinical benefit to ultimate outcome. Postmarketing studies would usually be studies already underway. When required to be conducted, such studies must also be adequate and well controlled. The applicant shall carry out any such studies with due diligence.

Linking a biomarker to a clinical endpoint is central to any discussion of biomarkers in clinical drug development. The process often involves establishing a significant statistical correlation and implies a causal or mechanistic association of the intervention with the disease process [2]. Linkage is often referred to as *validation*, a term also applied to performance characteristics of an assay or measurement, such as sensitivity, specificity, and precision (see Chap. 6). Because validation implies a linkage between the marker and clinical outcome that is not intervention or treatment specific, which is not always the case, the BDWG has recommended that the process of determining surrogate endpoint status be referred to as *evaluation*. Despite this legitimate distinction, validation and evaluation will be used interchangeably in this text.

III. BIOMARKER DISCIPLINES

Biomarker disciplines include therapeutic area, medical subspecialty, and disease-specific applications; specific technologies such as clinical pathology, imaging, electrophysiology, and applied genomics; comparative medicine, or species-specific challenges; principles of analytical and clinical validation; quality assurance and regulatory compliance; and research partnerships. With a focus on application of biomarkers to clinical drug development, we have included in this text chapters on these subjects as well as specific processes and concerns germane to this task. The latter includes dysregulated cardiac repolarization, which, together with the toxic potential for liver damage, represent the two greatest safety concerns of pharmaceutical sponsors and regulators; preclinical safety evaluation; the use of biomarkers as surrogate endpoints; applications to pharmacokinetic/pharmacodynamic modeling and clinical trial design; and the strategic application of biomarkers across the drug development process.

A. Laboratory-Based Biomarkers

Laboratory biomarker areas for discussion include the use of routine, or established, lab tests (including esoteric testing); novel applications of established lab tests; and the development and application of truly novel laboratory biomarkers. How these tests are developed, validated, applied, and the resultant data cleaned and interpreted arguably comprises the most important of the biomarker disciplines for the following reasons: (1) they, in aggregate, represent two-thirds of the objective clinical data in a typical NDA; (2) they are the means by which we systematically screen clinical trial subjects for organ toxicity; and (3) they are the most common markers employed to demonstrate drug activity and early indications of efficacy. Consequently, much attention is given in clinical drug development to operational issues relating to lab support, as well as venues through which special lab tests can be developed. The many challenges this presents pharmaceutical sponsors and business partners are reviewed in Chap. 2, with those related to clinical operations in Chap. 13.

B. Imaging

Imaging, as a biomarker discipline, is an eclectic group of markers that provide images of organs, functions, and a range of specific analytes that now obviate the need for invasive techniques requiring surgery or the insertion of needless tubes, wires, or catheters into the body. Imaging techniques include routine radiographs, ultrasonography, computed tomography (CT), magnetic resonance imaging (MRI), positron emission tomography (PET), single proton emission tomography (SPECT), and other scans that use isotopes to visualize and characterize organs, tissues, and pathophysiological processes that have been inaccessible prior to the availability of these technologies.

In drug development, these tools are now critical to demonstrating efficacy in many target diseases, including osteoporosis, arthritis, cancer, and vascular occlusive disorders. Pharmaceutical sponsors now spend tens of millions of dollars per year on resource- and data-intensive markers such as dual-energy x-ray absorptiometry (DXA) to measure bone mineral density, CT scans to demonstrate reduction in tumor size, and MRI to track the progression of osteoarthritis. Molecular imaging techniques include PET, SPECT, magnetic resonance spectroscopy, optical imaging, and infared fluoroscopy, and are now routinely employed in early-phase development to measure receptor occupancy using radiopharmaceuticals, as well as biodistribution in the whole body (absorption, distribution, metabolism, excretion) or specific organs such as brain. Accordingly, imaging can provide early indications of the activity and target selectivity of candidate drugs in early-phase development, or provide primary efficacy data in registration-phase clinical development. These applications will be discussed further in Chap. 3.

C. Electrocardiographic Monitoring

Because cardiac muscle, or mechanical function, is propagated and regulated by electrophysiological processes, the electrical state of cardiac myocytes is a logical marker to use to monitor this vital and vulnerable organ. This is accomplished through the surface electrocardiogram, or ECG, which represents the summation of the myocyte electrical activity [7]. The highly coordinated atrial and ventricular contractions are associated with synchronized negative and positive electrical potentials, which reflect polarization and depolarization, respectively. The return to a polarized state is called *repolarization.*

For the ventricle, which carries the heaviest burden for cardiac output, this process is reflected by the Q-T interval on the ECG, and is especially vulnerable to prolongation and dysregulation by pharmaceutical agents [8]. This can result in a pathophysiological state that can lead to serious, life-threatening ventricular arrhythmias, including a syndrome called torsade de pointes and sudden death. Q-T-interval prolongation is regarded as a biomarker for patients at risk for this rare and lethal cardiotoxic event. Accordingly, all candidate drugs are scrutinized at the preclinical and early-clinical phase of drug development for this activity. The importance of this electrophysiological marker and the effectiveness of the process used to monitor this activity, is underscored by the sobering fact that Q-T-interval prolongation and severe idiosyncratic hepatotoxicity are far and away the two most common safety issues that result in disapproval of NDAs and withdrawal of a drug from the market. Our understanding of the pathophysiology involved, the implications for patient care, and the measures taken to manage this risk in preclinical and clinical drug development are discussed in detail in Chap. 4.

D. Interspecies Biomarkers

Information derived from preclinical, or nonclinical, studies provides important biomarkers for clinical drug development. These diverse research activities include traditional biology and chemistry discovery efforts that identify, select, and characterize both targets and candidate molecules; pharmacology studies that define mechanism of action; and toxicology studies that employ in vitro and animal models for safety evaluation. The latter are particularly critical in selecting the candidate with the greatest chance of clinical success, through the determination of dose-limiting toxicities and other information used to define the margin of safety and, ultimately, the appropriate starting dose for clinical trials. Selection of the in vitro and in vivo models that are the most relevant to the therapeutic target and the human patient, and the design of studies that employ these tools, are important determinants of the predictive value these preclinical "biomarkers" hold for safety and efficacy in humans. Also important, therefore, are the biomarkers applied to these animal models—especially those used to determine the toxicological profile of the candidate drug. Having an informed perspective on species specificity and comparative medicine in applying the many novel and established biomarkers employed in a comprehensive preclinical safety evaluation is among the many challenges that our toxicologists and toxicological pathologists face today. How these are used to define toxicokinetics and the time- and dose-dependent association between drug exposure and biomarker alteration are, together with the previously mentioned preclinical challenges, discussed in Chap. 5.

E. Analytical Validation of Biomarkers

The credibility, utility, and ultimate value of biomarker data hinge on the validity of the assay or diagnostic procedure, as regards both the ability to measure the analyte or biological attribute and the linkage of that marker to a clinical feature or outcome. The former requires demonstrating and documenting performance characteristics such as accuracy, precision, linearity, specificity, sensitivity, range, reagent stability, and other characteristics of the test. The degree or rigor of analytical validation required of a biomarker used in drug development generally increases, in accordance with the stage of drug development, where the aforementioned features of a marker used to demonstrate target activity in a discovery laboratory will often be far less critical (and receive less regulatory scrutiny) than one applied to a stratified population of human subjects in a phase II or III clinical trial. While analytical validation is critical to the serviceability of any biomarker assessment, be it molecular imaging or a genetic marker, it poses the greatest and most common challenge in drug development for novel laboratory (in vitro) assays used to measure chemical, colorimetric, chromatographic, immunochemical, and cell-based endpoints. Despite the growing importance of biomarkers in drug development and the critical nature of appropriate stage-specific analytical validation, remarkably little guidance is available to drug hunters in this area. Opinion leaders in this technically challenging field have come together to offer such guidance in Chap. 6.

F. Clinical Validation of Biomarkers

As discussed previously, biomarkers are used diagnostically to define susceptibility to disease, disease progression, and response of the disease (clinical endpoint) to treatment. The first two categories of clinical biomarker applications cover broad disciplines and subspecialties relating to internal medicine and clinical pathology. Validation in this context relates to establishing the predictive value of a biomarker for a specific organ or systemic disorder. Factors that influence the predictive value of a clinical biomarker include sensitivity (number of false negatives), specificity (number of false positives), and the prevalence of the disorder in the population tested [9]. The importance of understanding the predictive value of routine and experimental biomarkers used to diagnose and manage disease cannot be overstated as a key determinant of effective health care delivery.

In clinical drug development, the focus of concern regarding validation relates specifically to defining the complex relationship among the biomarker, the clinical endpoint (disease target), and the effects of treatment on both. The kind of analyses required to fully characterize these associations are described in Chap. 7. The authors distinguish between defining the relationship between a particular marker and a clinical endpoint at the individual level and the effects of treatment on these measures at the clinical-trial level.

G.   Pharmacokinetic/Pharmacodynamic Modeling and Clinical Trial Simulation

The impact of biomarkers in drug development is defined by how they are used across the value chain of drug-target rationale, discovery, preclinical development, clinical development, regulatory approval, and labeling information. Perhaps the most important recent advance, as regards these broad applications, has been the emerging science of pharmacokinetic/pharmacodynamic (PK/PD) modeling and clinical trial simulation. It is now widely accepted that these critical tools improve the efficiency and productivity of these processes, by defining dose–concentration–effect relationships, and as a means for evaluating untested study designs, dose levels, and/or dosing regimens. These applications are discussed in Chap. 8 in the context of three highly challenging subdisciplines: mechanism-based models of disease, mechanism-based therapeutic interventions, and relationships between plasma drug concentrations and therapeutic and toxic responses.

H.   Genomic Biomarkers

The mapping of the human genome, together with the advent of technologies that enable practical and large-scale genotyping and screening for single nucleotide polymorphisms (SNP), has provided both a foundation for a medical science renaissance and a high-tech toolbox for drug discovery and development [10,11]. The contributions of these technologies to clinical drug development relate to their use in identifying new and highly specific clinical biomarkers. Genomic-based biomarkers useful in clinical drug development have two broad applications: as markers for disease susceptibility, and as pharmacogenomic endpoints. The former hold promise, as yet largely unrealized, as tools for drug discovery and clinical-trial patient stratification, where the latter offer immediate and substantial value in predicting drug metabolism, response to treatment, and toxicity. Such applications, however, require the same vigorous analytical and clinical validation to be serviceable in the highly regulated clinical trial setting. Ethical and legal concerns further complicate the challenging task of realizing the full potential of pharmacogenomics. These issues are explored in Chap. 9.

I.   Quality Assurance and Regulatory Compliance

Drug development is, in general, a highly regulated endeavor. The clinical development process is particularly scrutinized for compliance with the extensive regulations that ensure minimum quality standards and data integrity. It is therefore particularly curious that, given the growing importance of novel biomarkers in clinical drug development today, there are so few specific regulations that provide direction and guidance for how this vast array of technologies should be applied. Equally challenging, and only marginally more transparent, are the guidelines that direct sponsors regarding contract laboratories and collaborations with specialized biomarker service providers and academic institutions. Chap. 10 addresses these challenges and provides guidance as to how pharmaceutical sponsors can ensure that both internal and outsourced clinical biomarker research support is compliant with regulations and consistent with the principles of an effective quality management system.

J. Biomarker Research Partnerships

Partnerships in conducting the biomarker research required to meet the demands of clinical drug development have now become an essential component of pharmaceutical sponsors' clinical development strategies. This is driven by the increasing need for deep subspecialty biomarker research expertise, capacity issues, and operational requirements (data and specimen management capabilities, etc.) in which a sponsor may not choose to invest. The rationale behind such sourcing strategies for large-scale, routine contract biomarker support, such as central laboratories, ECG, and imaging, is obvious. Less so are the partnerships that support novel biomarker discovery, development, and validation.

Business principles and critical success factors for the former are discussed in Chap. 13. Strategies for more novel biomarker research partnerships are shaped by the complexity and rapidly changing nature of the clinical research environment; the disease-specific subspecialty expertise and technologies required for the development of new, well-validated markers; and similar capacity and resourcing considerations that influence routine testing sourcing decisions. These factors, as well as the special challenges that public–private partnerships present, are discussed in Chap. 12.

K. Clinical Operations

Finally, the application of established and novel biomarkers in drug development poses many *operational challenges*. This is particularly true in large multicenter clinical trials where the magnitude of the investment and logistics can be daunting. Accordingly, operational and organizational effectiveness has become critical to the successful execution of both early- and registration-phase biomarker strategies. The principles that underpin successful clinical operations units within pharmaceutical R&D organizations comprise an additional biomarker discipline, which is increasingly recognized as an important core competency.

Important areas for consideration include whether to centralize testing or the support within a pharmaceutical sponsor's clinical operations unit; the degree of standardization required to be successful; systems, or information technology support; staffing and use of consultants; and selection of service providers and business partnerships. These principles are discussed in the final chapter of this text, Chap. 13.

REFERENCES

1. NIH Definitions Working Group. Biomarkers and surrogate endpoints in clinical research: definitions and conceptual model. In *Biomarkers and Surrogate Endpoints: Clinical Research and Applications*; Downing, G.J., Ed.; Elsevier: Amsterdam, 2000; 1–9.
2. Biomarker Definitions Working Group. Biomarkers and surrogate endpoints: preferred definitions and conceptual framework. Clin. Pharmacol. Ther. **2001**, *69*, 89–95.
3. Mildvan, D.; Landay, A.; De Gruttola, V.; Machado, S.G.; Kagan, J. An approach to the validation of biomarkers for use in AIDS clinical trials. Clin. Infect. Dis. **1997**, *24*, 764–774.
4. Delmas, P.D. Bone marker nomenclature. Bone **2001**, *28*, 575–576.
5. SHEP Cooperative Research Group. Prevention of stroke by antihypertensive drug treatment in older persons with isolated systolic hypertension: final results of the systolic hypertension in the elderly program (SHEP). J. Am. Med. Assoc. **1991**, *265*, 3255–3264.
6. Food and Drug Administration Modernization Act of 1997, Title I, Subtitle B, Section 112; US Government Printing Office: Washington, DC.
7. Jalife, J.; Delmar, M.; Davidenko, J.M. Anumonwo, J.M.B. *Basic Cardiac Electrophysiology for the Clinician*; Futura: Armonk, NY, 1999; 1–38.
8. Singh, B.N.; Vaughan Williams, E.M. A third class of anti-arrhythmic action: effects on atrial and ventricular intracellular potentials, and other pharmacological actions on cardiac muscle. Br. J. Pharmacol. **1970**, *39*, 675–687.
9. Galen, R.S.; Gambino, S.R. *Beyond Normality: The Predictive Value and Efficiency of Medical Diagnosis*; Wiley: New York, 1975.
10. Lander, E.S.; Linton, L.M.; Birren, B.; Nusbaum, C. Initial sequencing and analysis of the human genome. Nature **2001**, *409*, 860–921.
11. The human genome: science genome map. Science **2001**, *291*, 1218.

2
The Clinical Laboratory and Collection of Biomarker Data

Gordon F. Kapke
Covance Central Laboratory Services, Indianapolis, Indiana, U.S.A.

Robert A. Dean
Lilly Research Laboratories and Indiana University School of Medicine, Indianapolis, Indiana, U.S.A.

I. INTRODUCTION

The clinical laboratory is an important research tool, providing a broad scope of biochemical, cellular, and morphological safety and efficacy biomarkers. While much of the biomarker testing is performed in the clinical laboratory, the processes provided by many laboratory facilities are well suited to support data generation performed at clinical study sites and other specialized testing facilities. When the biomarker measurements are performed on biological specimens, the consistent processes provided by the clinical laboratory help ensure an acceptably stable and appropriately identified sample for biomarker testing. A centralized clinical trial laboratory can minimize a broad spectrum of variables. Timely transportation and delivery of biological samples is just one example. These processes have become highly refined so as to provide for collection and analysis of specimens from most of our global geography. The audit trail available from a number of centralized laboratories can track collection and disposition of specimens and the availability of required laboratory and other biomarker data. By working with a central laboratory, clinical trial sponsors gain great expertise and efficiency in sample collection and specimen handling. Analytically robust testing is the hallmark for centralized facilities and, when based on consistent method principle and calibration, provides data that can be combined across time and geographies. The centralized clinical laboratory also employs a spectrum of operational processes designed for efficient collection, verification, and presentation of biomarker-related data. Timely access to clean, combinable biomarker data is a key component in efficient translation of data to information and knowledge. This chapter examines key preanalytical, analytical, and postanalytical biomarker-related requirements and central laboratory capabilities to address these requirements.

II. PREANALYTICAL CONSIDERATIONS

Preanalytical planning is critically important to the success of the development project. The inclusion/exclusion lab criteria are critical parameters defining the study population. Proper definition of the target population significantly impacts the success or failure of any study. Proper selection of tests and action limits will clearly minimize the risk of an individual with an inappropriate clinical condition being included in the test population. Key questions in determining the criteria for action limits of a biomarker are the diagnostic sensitivity, specificity, disease prevalence, and therefore the predictive value of a marker for a specified condition. Additionally, likelihood ratios may be beneficial to understand laboratory test limits to properly define a specified population [1].

Preanalytical sources of biomarker variability of interest are those that cloud the interpretation of the parameters reflecting normal biological function, the pathophysiological state or progression, and response to therapeutic or experimental intervention. Consistent and meaningful measurement of these primary parameters requires a thorough understanding of preanalytical variables that influence the biomarker measurements. Preanalytical biomarker issues include, but are not limited to, patient/subject preparation, specimen type and collection process, analyte stability, postcollection handling, transport, and collection of associated demographic information. Many of these preanalytical issues are impacted by the investigator site. Accordingly, efforts to adequately control the impact of these variables requires the cooperation of study site investigators and supporting staff. The central laboratory strives to ensure this collaborative cooperation by setting up routine and study-specific processes that are easy to execute. Whenever possible, the central laboratory attempts to address potential preanalytical issues through the design of processes that are transparent to the site. In well-designed processes, the site does not have the opportunity to act independently or inconsistently with the protocol-specified requirements related to preanalytical sources of variation.

The analytical validation process should precisely define the specimen type. For a given specimen type, many different collection containers are commercially available to an investigator. For instance, if one were to specify EDTA plasma as a biomarker sample, there are at least 25 different collection vacutainers, which yield EDTA plasma from whole blood. These collection tubes vary by EDTA concentration (15% or 7.5% K_3EDTA, yielding sample dilution of 2.4% or 1.1%, respectively, or spray-dried K_3EDTA, with no sample dilution), draw volume from 2.0 mL to 10.0 mL, size from 13×75 mm to 16×100 mm, glass or plastic, and stopper of the conventional and Hemogard design. With the opportunity for such diverse collection containers for an EDTA plasma specimen, the best opportunity for consistent sample collection is provided by constructing the collection materials into a well-designed protocol/visit-customized package. This ensures that the required collection containers appropriate for the planned analysis are utilized for the collection process and that the institutional review board (IRB)-approved volume of sample is obtained.

An additional characteristic of the collection container is the expiration date. Laboratory practice guidelines require that collection containers be used prior to their expiration date. Evacuated collection tubes frequently have a 6–12-month shelf life once they arrive in a laboratory environment. Although this seems like a substantial shelf life, one can frequently encounter delays of 6 months or more in the biomarker data collection process due to issues unrelated to laboratory operations. One should have a process to validate and document that the collection containers are within stability when the samples are collected. A well-documented laboratory collection process provides this audit trail and will produce rapid, definitive answers regarding the precise lot of collection containers, container expiration date, the manufacturer's product number, and the associated specifications of the product. This type of material audit trail is critically important when manufacturers of such devices change production processes, discontinue production, or recall or identify potential concerns regarding their product. By permitting a central laboratory to provide these collection materials, one is able to generate a detailed audit trail that helps answer questions that might arise. The part numbers of the collection supplies along with the lot numbers are recommended to be a permanent record of every sample for the project. An appropriate audit trail of the shipment of containers and their associated product identification numbers and lot numbers to investigators should be generated and retained to document all movement and possession of the sampling supplies.

Once the required collection container has been precisely specified, the collection process needs to be defined in detail. Is the sample to be collected using a traditional venipuncture needle, a butterfly needle, or through a heparin lock? The collection process will vary depending upon the patient population and the study design. The goal for successful data collection is to have the proper devices matched to the patient at the point of collection. A central clinical laboratory has the experience to help in the design of the collection process dependent on the patient population, the setting for specimen collection, and the requirements of the biomarker to be measured. Collection instructions need to be customized for the planned patient encounter. Specifically, the collection instructions must address critically important biological issues for the specific biomarker(s) to be measured. For example, the fasting state of the patient and duration of that state, patient positioning [supine or erect prior to and/or during specimen collection— fluid redistribution within the patient going from erect to supine can result in more than a 5% decrease in some measured biomarker concentrations [2]], and timing of specimen collection (first morning, second morning, or random urine collection) often need to be defined. As noted above, the laboratory can provide the exact collection materials but can also provide detailed processes and procedures for the investigator. Providing very specific project instructions is important as many investigator sites have multiple studies ongoing at a given time. It is important that the study instructions be printed in languages that are easily understood by investigator study site personnel. Specific collection instructions should include the order of collection for different vacutainer types as the vacutainer anticoagulants can cross-contaminate specimens. The site also needs to be sensitive to incomplete vacutainer fills (short draws). Short draws are a concern for all anticoagulated samples. The most sensitive sample to a short draw is citrated plasma for coagulation assays. Short draws change the anticoagulant to plasma ratio, alter the sample dilution (citrated plasma is diluted by 10%), and result in an invalid sample.

Following specimen collection, the sample must be handled in accordance with the biomarker validation. It is essential that site personnel be provided readily available sample-handling instructions. The clinical laboratory can provide collection instructions maximizing the opportunity for proper specimen collection and handling. These instructions should specify the maximal time delay in sample processing, the time, temperature, and force of centrifugation, as well as any additional sample-processing instructions. The clinical laboratory also can provide transport containers, transport outer packaging, courier information, and airbill. Preprinted courier airbills addressed to the receiving laboratory facilitate reliable transport of specimens between the investigator sites and the testing location. By working through a clinical laboratory, the sponsor can document specimen movement from the point of collection to the testing location. Couriers generally perform well; however, all critical sample shipments need to be tracked to their destination as delays do occur and biological specimens are time-sensitive shipments. Delay in transit can result in specimens being received in a degraded condition. Effective tracking can allow for interventions like adding additional dry ice to delayed shipments.

Analytical data are of value only if correctly associated with demographic data. The clinical laboratory has experience and systems to ensure that the essential demographic data are correctly captured. Collection devices and transport containers need to be linked to the patient in a robust manner. Transport vials are optimally labeled with a primary and secondary identifier. The identifiers need to be linked to collection information such as the time and date, the specimen was obtained and the master patient record. The clinical laboratory can provide customized requisitions for efficient capture of this demographic and collection-process-related data. Additionally, the laboratory clinical trial information system can easily capture and maintain this record. Careful preplanning of the data flow by the study director can ensure that the processes adequately address the needs of the study. Those responsible for data management need to be involved in the planning process before the predefined database is constructed. The proper naming of visits is as important as generating the data. Without careful coordination, the project team may generate much information that is difficult to transform into knowledge because data elements cannot be efficiently segregated in a manner that renders project data most useful.

The central laboratory acts as the link between the investigator site and the biomarker testing laboratories. Accordingly, the project sponsor should have a mechanism to verify the specimen identification against a central database. Since specimen identification is a critical site activity prone to inconsistencies, an intelligent clinical trial system will predefine visits attached to the investigator and be capable of determining whether a patient is participating in the appropriate visit. Additionally, the system should be capable of matching patient demographic information against a master patient file because missing demographics or clerical inconsistencies in demographics are a common finding. By resolving these demographic inconsistencies shortly after specimen collection, one has the greatest opportunity to correctly clarify any demographic questions. Resolution of the demographics issues early in the project process ensures the generation of a clean project database that can facilitate rapid data transmission, analysis, and knowledge creation.

The site is a key participant in the biomarker data collection and also the most unpredictable variable in the data collection process. Detailed training of investigator sites typically yields a positive return on investment. Our experience has been that the site is the primary party generating inconsistencies in demographic data. These issues with demographic data are easily identified using electronic checks applied to the data and can be substantially reduced with site training. We have experienced poor study compliance in studies based on protocols deemed to be "easy" and not thought to require formal site training. By contrast, we have participated in very complex protocols where it was clearly recognized that sites needed to be motivated and well trained. Attention to site-training issues resulted in successful on-time completion of these operationally challenging protocols. Processes available from the central laboratory can track the performance of each study site in collection of patient demographic data and site-associated testing cancellations. This approach can help target the need for focused retraining of study sites when needed. Site performance information may also be useful in determining future study participants.

The biomarker validation will document specimen stability. However, issues of biomarker stability are somewhat different from stability issues for a new chemical entity. With a biomarker, highly purified reference material is often unavailable for preparation of "spiked" control specimens used to characterize analyte stability. Some biomarkers, for example those that are cell-associated, cannot be prepared and stabilized by freezing, as is done with biochemical analytes measured in serum or plasma. The collection of biomarkers from multiple sites, often from diverse geographies, requires detailed assessments of procedures for specimen handling, shipping, and storage. The central laboratory can assist the sponsor in assessing analytical method–related concerns.

With planning, a central laboratory can facilitate biomarker collection any place in the world. Existing courier networks can move specimens from many locations in the world to a second location within 48 h. Specimen stability mandates the method and condition of transport. Transport of specimens under ambient conditions is the most cost-effective. However, the extremes in ambient conditions can vary dramatically by season and location. The central laboratory can limit the extremes in ambient conditions through use of specialized packing materials. Similarly, refrigerated conditions ($2-15°C$) can be maintained during transport for 48 h. Materials for refrigerated shipping are more costly than those for ambient shipping. The cost for air transport of refrigerated specimens is comparable to that of ambient specimens on a dimensional weight basis and does not require use of hazardous materials. Shipment of frozen specimens is substantially more costly than that for ambient and refrigerated specimens. Dry ice is a hazardous material requiring special freight handling for air transport. As a result, the flight captain of passenger aircraft can deny boarding of dry ice and other aviation hazardous materials. The limited access to air transport for frozen specimens and other hazardous materials limits substantially specimen shipping options. One needs to be sensitive to the biohazard classification of the transported material. Known infectious material must be declared as infectious, as these materials are classified as hazardous for air transport.

Different forms of hazardous materials may be required to generate biomarker data. These hazardous goods range from dry ice, to infectious specimens (HIV, HBV, HVC, etc.) and noxious and flammable chemicals such as tissue fixatives [3]. The volume of material transported is a determinant in the definition of aviation hazardous chemical liquid. A 10-mL vial of fixative transported separately may not be classified as aviation hazardous based on volume. A bulk supply shipment of the same fixative vials, owing to the total volume in the package, may be classified as an aviation hazardous material. The International Air Transport Association (IATA) Dangerous Goods Regulations also impose limitations on the quantity of infectious material per package that can be shipped on passenger aircraft. Hazardous material shipments significantly impact development budgets. Therefore, the development of specimen requirements and handling procedures to conduct biomarker data collection should be designed to minimize or, if possible, eliminate shipping and handling-related hazardous materials. Excellent logistical planning has the added benefit of controlling costs and avoiding undesirable delays in specimen transport.

In the process of biomarker assay validation, it is important to define conditions that stabilize the specimen and, if possible, permit ambient or refrigerated shipping and handling. For example, the stability of hemoglobin A1c (HbA1c) in EDTA whole blood was shown to be acceptable over 3 days at ambient conditions (20–25°C). HbA1c is also stable at − 70°C for at least 3 months. However, at − 20°C, as much as 10% of HbA1c was lost after 1 week (Fig. 1). One could ship specimens for HbA1c analyses frozen on dry ice, which maintains the temperature at approximately − 80°C. However, one does not want to delay the generation of this biomarker data by delaying shipment while dry ice is obtained, risk site storage of the specimen at − 20°C or less, and incur the high costs associated with dry ice shipments. Alternatively, having the site transfer a capillary of the specimen into the Bio-Rad preservative can stabilize the specimen for relatively long-term ambient or refrigerated handling [4]. Once a specimen is opened and processed at the site, the risk of improper identification

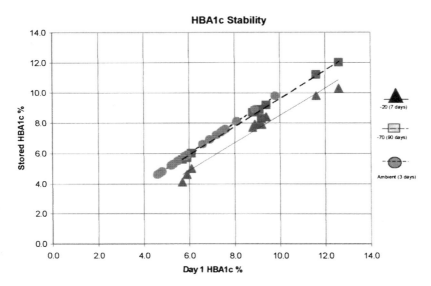

Figure 1 Patient specimens were transported at ambient conditions to Covance Central Laboratory and analyzed the day after collection and additional analyses were performed after storage at the specified times and temperatures.

increases. Accordingly, key strategies for biomarker specimen collection involve maximizing ambient shipping conditions and minimizing the site requirements for intervention in the preparation of specimens.

Local laws regulating the movement of biological specimens following collection must also be considered. Although one can obtain specimens from most countries in real time, some countries will not permit real-time shipment of biological specimens across their borders. In addition, some countries prohibit shipment of specific sample types such as whole blood. Even when transport permits can be obtained, one must be careful to understand the sponsor's responsibility for obtaining these permits versus that of the laboratory.

In many countries, the sponsor is required to obtain a study-specific permit. In other countries, a permit is required for each individual shipment. Consult carefully with your central laboratory before starting a study to ensure that the sponsor obtains appropriate permits and that collected samples can be moved as planned for the project. If the biomarker is frozen, and has acceptable frozen stability, these custom restrictions can be managed. However, if a specimen has a limited stability, and cannot be appropriately stabilized, biomarker data generation may not be possible or may not be possible in the desired specialty laboratory.

Stability is a key parameter in collection of biomarker data. A study manager needs to understand the stability profile for an analyte and approaches for monitoring specimen stability throughout the project. Clinical laboratories have procedures for evaluating and documenting specimen stability from collection to analysis. These tracking processes can be completely manual or fully automated. The clinical laboratory can advise sponsors regarding the monitoring process or processes appropriate for a specified study. The clinical laboratory also is uniquely suited to deal with relatively unstable specimens required for biomarker data collection. While stability of some specimens cannot be extended, the 24-h per-day, 7-day-per-week operation of a clinical laboratory can facilitate specimen receipt at any time couriers can provide freight delivery. In unusual situations, a sponsor may choose a premium freight courier that will deliver night and day to the clinical laboratory. Once a specimen is in the laboratory, it can be analyzed to complete the biomarker data collection process or stabilized in manners that are not feasible at the collection site. The personnel of central laboratories have the experience and access to facilities required for safe use of hazardous materials for preserving specimens pending biomarker analysis.

Following validated processes for a particular specimen type does not ensure that a specific clinical specimen will be satisfactory to yield a valid result. Specimen collection by multiple sites and from multiple study patients will produce conditions not found in the validation environment. One of the more common concerns is the impact of various sample matrix constituents.

For example, the analytical measurement may be altered by endogenous compounds found in human plasma. Bilirubin, lipids, and hemoglobin are three of the most commonly studied endogenous substances affecting clinical specimen analysis. The biomarker may also be released, generated, activated, or inactivated by components of the human blood. For instance, Factor VII is activated in vitro by improper sample processing [5]. If the study site fails to promptly separate plasma from cells and platelets and rapidly freeze the sample, the Factor VII activity result may be very high and not represent the biological state of the patient at the time of specimen collection. Correspondingly, some biomarkers are degraded by intracelluar or extracellular components of blood. These sample integrity issues may not be recognized in the validation process. An example of analyte degradation is the loss of insulin in hemolyzed specimens. Insulin is generally considered to have a 12-h ambient stability [6]. If one collects the specimen with no visible hemolysis, 12-h stability can be achieved. However, hemolysis of red blood cells releases the enzyme insulinase, which degrades insulin. Even if a hemolyzed specimen is centrifuged, separated from the cellular material, and frozen in a short period of time, substantial loss (greater than 30%) of insulin can occur in the sample [7]. With the rapid proliferation of new biomarkers, similar unknown biological preanalytical findings are highly probable. In addition, one faces a multitude of exogenous materials that will be present in the patient population and frequently not present in validation samples to include combinations of drugs, nutraceuticals, and food-related compounds. Accordingly, it is important to examine unusual or nonroutine specimen collection and handling conditions and a broad spectrum of specimens from the population under investigation to identify the potential impact of these concerns. Literature summaries to help navigate these issues are available [8–10].

In the process of test validation, biological variation should be explored. The intraindividual biological variation may have a significant seasonal, daily, or monthly component. The difference between a fasting and a nonfasting state may be significant. Common challenges in development of novel biomarkers are the lack of ready access to specimens from relevant populations and the limited time available to document potential issues. For known biomarkers, a thorough literature review and appropriately focused validation experiments can help control or minimize the impact of biological variability that might confound experimental interpretation of biomarker data. A detailed discussion of biological variation is available in the literature [11].

For biomarkers that are widely accepted in clinical practice, it is highly recommended that the laboratory facility evaluate analytical testing by participating in a commercially available external proficiency program. Although the proficiency testing programs often use specimens prepared from an altered specimen matrix, the comparative data derived on "blind" analysis of these specimens provide valuable objective information about ongoing method performance. The data generated by a clinical laboratory for a particular routine, commercially available assay can be compared to data derived by other laboratories using the same or similar analytical platforms and reagents. In some cases, proficiency programs provide an accuracy base linked to definitive analytical methods. At the very least, these programs provide a basis for comparison of biomarker data between laboratories. Additionally, external blind proficiency challenges are a convenient means of demonstrating assay performance to regulatory organizations.

Clinical and scientific organizations certify the analytical performance of laboratories for biomarker methods of pivotal concern to research interests of the organizations. If available, certification of a biomarker assay by such an expert group is recommended. For example, the Centers for Disease Control (CDC-NHLBI Lipid Standardization Program) provides laboratory certification for measurement of serum cholesterol, triglyceride, and HDL cholesterol. Similarly, the National Glycohemoglobin Standardization Program (NGSP) certifies hemoglobin A1c assay performance. What do these certification programs provide? First, like other proficiency programs, these programs provide an external, independent assessment of analytical accuracy and precision. This is important, as internal assessments of analytical accuracy are typically based on internally prepared or commercially purchased reference specimens. Errors in the preparation and analysis of these materials can and do occur. These external programs provide a means to identify and correct such errors. Second, clinical laboratories have access to a host of analytical methods for a single analyte. In the case of HbA1c, at least 17 methods are commercially available. The NGSP certification is available for more than 11 methods at this time (Table 1). A current list of NGSP certified methods can be located on the NGSP website (http://web.missouri.edu/~diabetus/gnsp.html). The accuracy performance among the commercially available methods, however, is not the same. Among the NGSP-certified methods, as demonstrated by the College of American Pathologists (CAP) survey results GH2-01, 02, and 03 from the first survey of 2002, the maximum systematic difference (span low to high) observed among methods is 0.8% at a HbA1c reference value of 5.2%, 0.9% at a reference value of 8.0%, and 1.5% at a reference value of 10.9%. For all the certified methods, the method-specific medians were all within 0.4%, 0.5%, and 0.9% of the reference values of 5.2%, 8.0%, and 10.9%, respectively. The HbA1c assay from a single manufacturer but of different assay versions varied 0.4% at a reference value of 5.2% and 0.6% at a reference value of 10.9%. This within-manufacturer bias was 7.6% and 5.5% at reference values of 5.2% and 10.9%, respectively. Inattention to and failure to control this magnitude of bias through use of a single laboratory committed to a specific method, reagent, and calibration method could confound the interpretation of some clinical investigations presumably based on analysis of the same analyte. External certification improves standardization. However,

Table 1 Comparison of CAP GH2 2002 HbA1c Survey Results by Method[a]

NGSP reference value[b]		CAP sample GH2-01		CAP sample GH2-02		CAP sample GH2-03	
		8.00		5.20		10.90	
%	No. labs	Median	%CV	Median	%CV	Median	%CV
Method 1	201	8.0	3.7	5.1	4.4	10.8	3.6
Method 2	260	8.2	7.3	5.5	8.0	10.9	6.1
Method 3[c]	123	7.6	3.6	4.8	5.0	10.3	4.3
Method 4[c]	76	8.1	2.6	5.2	3.2	10.9	2.3
Method 5[c]	287	8.1	3.4	5.1	3.2	11.1	3.8
Method 6	109	7.9	4.3	5.5	3.6	11.2	3.3
Method 7	24	8.0	5.7	5.3	7.0	10.8	6.4
Method 8	160	8.5	5.0	5.2	5.1	11.8	4.7
Method 9	147	8.0	5.3	5.6	5.3	11.3	5.8
Method 10	342	8.5	2.8	5.0	4.2	11.8	2.7
Method 11	9	8.4	-	4.9	-	11.7	-
Method 12	248	7.6	7.5	5.4	7.5	10.1	7.6
Method 13	12	7.6	9.7	4.5	13.9	11.0	8.6

[a] Methods 1–11 are NGSP certified.
[b] Assigned as the mean value of six replicate analyses over 2 days using four NGSP-certified secondary reference methods.
[c] Single manufacturer.
Source: Data obtained from NGSP website and used with the permission of the National Glycohemoglobin Standardization Program and the College of American Pathologists.

obtaining data from multiple certified methods does not ensure that data maintain maximum consistency. In the case of a biomarker such as HbA1c, an absolute bias in data collected across multiple testing platforms may mask a therapeutic effect or inappropriately support the expected outcome. Finally, the certification program provides a network of analytical facilities to advance clinical research dependent upon key biomarkers.

While well-organized programs do not exist for most novel biomarkers, many laboratories supporting clinical research cooperate to evaluate nonroutine or novel biomarkers. The comparison process is often accomplished through round-robin exercises involving the exchange of method comparison specimens. These cooperative approaches have been facilitated by biomarker focus groups recently organized with support from the Food and Drug Administration, the National Institutes of Health, academia, and the pharmaceutical and biotechnology industries (see Chap. 12).

III. ANALYTICAL CONSIDERATIONS

Lee, Smith, Nordblom, and Bowsher thoroughly review the analytical validation
of biomarker methods in Chap. 6 of this text. We will not readdress those issues.
There are, however, numerous analytical issues not routinely addressed in formal
experimental validation. Examples of issues that may not be recognized at the
time of validation are product availability, product life-cycle improvements, and
manufacturer's product release criteria.

At the analytical validation stage, one needs to understand the requirements
of assay support after registration. If assay support is required for drug
commercialization, one needs to develop that strategy during the development
program. The advantage of using commercially available methods to assess
safety or efficacy of administered agents is the availability of the same methods
for monitoring the effects of these agents upon approval and commercialization.
If a biomarker will be required for patient management after compound
commercialization, it is essential that the sponsor plan for an assay that will be
commercially available to the patient's physician at drug approval.

When developing or selecting a method, review the viability and quality
record of the supplier. Have they been able to respond to rapid shifts in market
demand? Do they have reliable raw material sources for the assay? Are they
willing to commit to maintaining the assay for the length of your development
program? Are they willing to supply you with the current assay version even if
they upgrade their general product for the laboratory industry? One needs to be
sensitive to these questions as one counts on consistent product availability once
committed to the assay validation. The current history of the diagnostic
companies is that these questions are not part of the typical product order
fulfillment process. For successful biomarker data collection, one needs to strive
for a long-term supplier relationship with the commitment to provide a consistent
product.

The diagnostic industry responds to change in the technology and
competitive pressures. Consequently, a product that has been validated for
biomarker data collection can be upgraded by the product manufacturer. If a
competitor improves the lower limit of quantification of the assay, the
manufacturer may be forced to follow the competition to maintain market share.
The manufacturer may be striving to gain market share from a competitor by
improving the sensitivity of an assay such as a qualitative assay for a viral
marker. The manufacturer may want to automate the assay and thereby improve
its competitive position relative to the cost of analysis for the performing
laboratory. All these issues will drive method changes and are required for the
survival of the assay manufacturer. However, all these competitive pressures
create difficulties when collecting biomarker data for research purposes.

In the biomarker testing process, there are unique standardization issues to be faced. In the assay of a study drug, one typically has a pure homogeneous material that can be utilized as a standard and recovery can be documented in a human matrix. With a novel biomarker, widely accepted purified reference material for preparation of standards may not be available. Accordingly, comparable method characterization experiments cannot be executed. One approach to understanding and documenting the long-term standardization of the method is to assay a consistently homogeneous population. If the population is well defined and reasonably stable, additional samples can be collected from this population at future dates to demonstrate that the numerical description of this population (mean, median, standard deviation, 95% limits) has not significantly changed over time. The consistency of the numerical population description is evidence that the standardization of the biomarker assay has not changed. Documenting consistent performance between laboratories increases the confidence in the ability to properly generate the biomarker data. In addition, without the benefit of pure standards, longitudinal comparison of banked population samples provides evidence of testing consistency. The use of all laboratory-generated patient data from the population is an additional means of documenting consistent long-term assay performance. Documentation of the performance of the "average of normals" in the clinical laboratory has been used to demonstrate consistent assay performance and has been utilized as a quality control process [12]. The "average of normals" concept has been demonstrated to be beneficial in the routine clinical laboratory and has the same potential in a biomarker laboratory with appropriate knowledge of the dynamics of the population of samples being analyzed.

Even when appropriate standards are available one needs to be sensitive to the manufacturer's tolerances in standard assignment. Frequently we have observed that the manufacturer does not use a true human specimen to assess the accuracy of the calibration assignment. For example, we recently noted a 50% drop in the recovery value of an esoteric analyte in longitudinal human pools used to quality-control the parameter. The comment to date from the manufacturer is that the product passed manufacturing specifications, which appears to mean the standard curve has the expected signal for the specified standard concentrations. No recovery studies appear to be have been performed on human specimens prior to lot release by the manufacturer.

Some manufacturers have product release specifications that are inappropriately broad based on the total error specification expected for the analyte. Our experience is that a manufacturer had a $\pm 3.0\%$ specification on the calibrator value assignment for a specific analyte. The clinical total error specification for this analyte is 1.5% with less than 1% being allocated for a bias. Total error is defined as: $TE\% = Bias\% + 1.65 \times CV\%$. The manufacturer's acceptable calibration bias is greater than the laboratory expectation for total error. The clinical laboratory needs to understand all the operating characteristics of the biomarker reagent manufacture as these characteristics frequently can be inconsistent with the intended experimental use of the biomarker. In addition, a laboratory performing a high volume of analyses will generate much more quality-control data of a product and thereby better understand the performance of the product than does the manufacturer. Effective partnership between the clinical laboratory and the manufacturer can help to swiftly and effectively identify and correct product performance issues.

Even with the best intentions of the manufacturer, issues can arise where consistent performance of an assay cannot be maintained. One needs a close working relationship with the manufacturer to work through real and perceived issues. Ultimately the manufacturer may have to change the operating characteristics of the product, and the testing laboratory and data analysis group will have to adjust to the change. For example, after a manufacturer gains experience with multiple lots of raw materials required for reagent manufacture, it is not uncommon that a performance characteristics of an assay, like the lower limit of quantitation, must be adjusted. The laboratory and the sponsor need to understand these issues and develop strategies to address these problems in a manner that does not compromise the conduct of the investigation. The best-prepared plans may need to be altered.

Data generation in real time is recommended. By generating biomarker data continuously one can assess site and lab performance. If one waits until the end of the study, one may find specimens have been collected in a manner that does not permit generation of valid analytical data. Coagulation biomarkers are a good example. Coagulation samples are very sensitive to collection artifacts such as coagulation pathway activation in the collection process and therefore fibrin generation before freezing. Laboratories will cancel coagulation samples containing detectable fibrin clots [5]. If one is not analyzing the samples during the study, one may find that the yield of samples is very low owing to difficult collection issues. From a patient population basis, pediatric sample collection is a prime area to face these collection-induced sample issues. From an analyte perspective, coagulation is very sensitive to preanalytical sample handling procedures. Batch analysis will produce the minimal between-run analytical variation; however, the increased data consistency needs to be weighed against the operational factors in determining how to best generate data. Ideally, data will be generated with specimens flowing through the analytical procedure in a random manner thereby minimizing the effects of reagent, calibration, and between-run bias. Alternatively, if the specimens are stable for batch testing, all specimens may be tested in a short period with the same lot number of reagents thereby minimizing reagent bias and maximizing data consistency. For biomarkers with a short stability like coagulation factors, coordinated planning with the reagent manufacturer will permit minimum lots and maximum lot expiration dating thereby maximizing data consistency.

IV. POSTANALYTICAL CONSIDERATIONS

Upon completion of the analytical task, data must be delivered to a number of end users for analysis, interpretation, and action. Blinded and unblinded personnel need to be clearly identified to ensure that project personnel have access only to the data they are authorized to receive. This requires a process and infrastructure designed to efficiently move data from the location(s) performing the analyses to the personnel accountable for patient care and operational and scientific execution of the trial. The information system is the key instrument providing timely access to data formatted to the needs of these end users. The reporting processes of the information system need to be sufficiently flexible to address the requirements of the various users of the data. Personnel responsible for patient care need information organized in a way that helps them effectively evaluate the status of individuals in their care. By contrast, the protocol scientific personnel typically need study datasets so that they can evaluate drug response within and/or between study populations or population subsets.

The clinical laboratory can provide real-time notification to the patient care personnel as well as the project personnel from the test action limits that are built into the clinical trial laboratory database through automated and manual processes. The laboratory must assume responsibility for communication with the patient care personnel as outlined in the protocol and supporting documents. In a high-throughput automated facility, the laboratory must have a process where critical patient safety information is conveyed directly to the caregiver with positive confirmation that the information was received. By providing real-time feedback through the reporting process, the clinical laboratory becomes an essential communication link in successful biomarker data delivery. The reliable delivery of laboratory test results and associated flagged results is a critical element ensuring patient/subject safety. The central laboratory with congruent global databases is capable and critically important in executing these identification and notification processes for large multisite, multinational global trials.

The laboratory processes should be capable of rapidly resolving any patient identification issues so reporting of critical patient care results and reporting of important protocol actions parameters will occur in a timely manner. A highly predefined protocol database allows the central laboratory to identify the investigator site and notify the site of critical patient care results even when demographic information is in question. The laboratory processes also should have backup strategies to cope with the lack of a response by the investigator.

This may involve engagement of the laboratory director and key personnel from the study sponsor.

Sharing of data with the investigator is a critical element in patient care and safety. The laboratory can generate, organize, and communicate much of the data needed by the investigator. The laboratory also can identify and "flag" laboratory and other biomarker findings that are notable and may require a specific response on the part of the investigator site. These central laboratory services are optimally managed through use of a predefined, project-specific database with automation of the required functionality. Well-designed laboratory information systems can communicate data, related information, and the need for specific follow-up to any location in the world. This capability enables the investigator to more efficiently interpret the laboratory and other biomarker data in the clinical context. Optimal interpretation of results in the context of an individual patient requires an understanding of the diagnostic sensitivity, specificity, and predictive value of a given laboratory test [13]. An abnormal laboratory test result does not confirm the presence of disease or a clinical condition. Similarly, a normal laboratory test result does not rule out disease. The sensitivity and specificity of the assay will change with the absolute value of the reported results. The plot of sensitivity versus specificity is known as a receiver operator curve (ROC). A significant literature has been published on the use of ROCs to optimize the diagnostic limits of laboratory and other biomarker tests [13]. In addition, the predictive value of a laboratory test depends on the prevalence of the disease in a specific population. Owing to the interrelation of these parameters, one must understand that clinical laboratory findings should not, in and of themselves, be interpreted as adverse events. For this reason, the laboratory needs to partner with the physician investigator and the sponsor to ensure patient safety.

The data collected and reported by the laboratory extend beyond test results and alert flags associated with the results. The laboratory, by nature of its mission and the time-sensitive nature of the specimens, often has a very high level of site interaction during a clinical trial. As a result, the laboratory database can be used to track patient enrollment and progression of the patient through the study. As an example, the laboratory database can be used to monitor and compare the impact of the laboratory, the investigator site, and courier performance on screening test cancellations. Table 2 summarizes the reasons for the cancellation of laboratory tests per million requested tests. These global performance metrics were gathered by Covance Central Laboratory. The total testing volume is reported to give relative size comparisons among the testing locations. The categories of "lost in the laboratory" and "mishandled by the laboratory" are performance parameters directly linked to the analytical facility. Test lost due to use of "expired collection container," "hemolyzed specimens," and "WBC degeneration" are parameters related to site performance. Site and courier performance impacts the categories "broken in transit" and "received beyond stability." Tests not performed within stability are a combination of the laboratory processes and investigator site. Tests lost due to whole-blood microclots is an evaluation of specimen transport conditions. Inappropriate cooling or heating during transport, in addition to inadequate sample mixing at collection, can generate microclots in EDTA whole-blood specimens. The most common reason for lost data is specimen received beyond stability. The next two most common reasons for loss of data are specimen hemolysis and use of an expired collection container, which are directly linked to site performance. The typical data yield will be greater than 98.5% and direct laboratory causes for lost data are less than 0.1% of the total data. Regardless of the cause, this approach allows the drug development team to track such losses and identify the need for corrective measures should these losses prove unacceptable.

The laboratory also generates biomarker data needed by the protocol scientific personnel. Timely access to well-organized and verified datasets is key to efficient analysis, knowledge creation, presentation, regulatory filing, and decision making. This requires rapid availability of data in electronic formats customized to the specific demands of the individuals responsible for data analysis and interpretation. The clinical laboratory can customize data formats in many ways. A data transfer agreement is a written agreement between the central laboratory and the data recipient. This document allows various parties involved in the trial to define and prepare for the specific data requirements. Without a written agreement, data transfer requirements tend to continuously evolve. These uncertainties introduce timely and costly delays in the data delivery process. It is essential that key individuals generating, managing, and working with the data have an opportunity to be involved in the planning of the project.

Our involvement in and review of numerous projects has revealed the critical importance of collecting laboratory and other biomarker data that are combinable (same method principle, same calibration) over multiple studies supporting a drug development effort. The generation of combinable data improves opportunities to detect medical and statistical significance. Clinically insignificant analytical drifts and shifts can occur with methodological changes and preclude or confound the data analysis. Statistical significance is a function of population dispersion, magnitude of treatment effect, and the number of observations. The magnitude of analytical shifts and drifts permitted in a biomarker dataset for nonbiological changes needs to be minimized to appropriately define populations by the inclusion and exclusion criteria and to maximize the probability of detecting statistical significance among treatment groups. Literature has been published proposing acceptable total error in clinical laboratory measurements [14]. Owing to the challenge of demonstrating statistical significance in the biomarker data, the analytical procedure should always minimize controllable variation beyond the expectations of the routine clinical laboratory.

Table 2 Lab Performance Metrics

	Indianapolis	Geneva	Cape Town	Sydney	Singapore
Test count	1,506,943	668,268	36,124	46,508	32,920
Tests lost due to specimen lost in lab	54 ppm	34 ppm	9 ppm	179 ppm	0 ppm
Tests lost due to tube broken in transit	94 ppm	142 ppm	0 ppm	106 ppm	69 ppm
Tests lost due to specimen received beyond stability	5,071 ppm	10,018 ppm	2,441 ppm	2,764 ppm	4,856 ppm
Tests lost due to expired collection container	1,091 ppm	496 ppm	998 ppm	1,649 ppm	60 ppm
Tests lost due to test not performed within stability period	139 ppm	197 ppm	452 ppm	115 ppm	0 ppm
Tests lost due to mishandling by laboratory	96 ppm	435 ppm	9 ppm	41 ppm	17 ppm
Tests lost due to hemolyzed specimen	1,132 ppm	1,731 ppm	1,758 ppm	397 ppm	884 ppm
Tests lost due to microclot specimen	331 ppm	178 ppm	135 ppm	371 ppm	638 ppm

V. CONCLUDING REMARKS

Biomarker data collection is logically associated with preanalytical, analytical, and postanalytical phases. Each phase has critical elements that require careful planning and management to ensure a successful drug development project. The sophistication of the planning and management of these phases is becoming more challenging as we employ more and more novel biomarkers and novel applications to otherwise routine biomarkers. The central laboratory has considerable experience in the validation and application of novel analytical methods in clinical research. Accordingly, the central laboratory can be a valuable partner in the development project. The tools and techniques developed by central laboratories to support clinical research are increasingly being modified to better and more broadly support a host of nonlaboratory biomarkers. Examples include data derived on physiological (vital signs and pulmonary function tests), electrophysiological (electrocardiograms), and imaging-based measures. As the breadth of application support expands, the basic lessons with laboratory-based biomarkers hold true: plan well; train well; keep it simple; clearly exchange requirements and capabilities; and jointly perform a risk assessment. Finally, drug development is research, and the unexpected will happen. Accordingly, be prepared to identify and resolve unanticipated issues. The key to rapid development is understanding risks and having a means to manage through issues rapidly. The central clinical laboratory brings substantial experience and capability to partner effectively in biomarker-based development projects.

REFERENCES

1.   van der Helm, H.J.; Hische, E.A.H. Application of Bayes's theorem to results of quantitative clinical chemical determinations. Clin. Chem. **1979**, *25*, 958–988.
2.   Henry, J.B. *Clinical Diagnosis and Management by Laboratory Methods*, 16th Ed.; WB Saunders: Philadelphia, 1979; 12.
3.   Dangerous Goods Regulations (IATA Resolution 618, Attachment "A"), 43rd Ed.; International Air Transport Association: Montreal, 2000.
4.   Becker, S.J.; Maleki, S.H.; Graves, M.S.; Khan, M.H. Evaluation of frozen stability for hemoglobin A7c on the Bio-Rad variant classic using the Bio-Rad capillary collection system. Clin. Chem. **2000**, *46*, A39.
5.   Jensen, R. Preanalytical variables in the coagulation laboratory. Clin. Hemostas. Rev. **2001**, *15*, 1–4.
6.   Jacobs, D.S. *Laboratory Test Handbook*, 4th Ed.; Lexi-Comp: Hudson (Cleveland) OH, 1994; 149–150.

7. Pasic, J.; Bhatnagar, M.K.; Pickup, J.C. Self-collection by diabetic patients of capillary blood for free insulin monitoring: reduction by diamide of haemolysis-induced insulin loss. Diabetic Med. **1991**, *8*, 140–145.

8. Young, D.S. *Effects of Preanalytical Variables on Clinical Laboratory Tests*, 2nd Ed.; AACC Press: Washington, DC, 1997.

9. Friedman, R.B.; Young, D.S. *Effects of Disease on Clinical Laboratory Tests*, 3rd Ed.; AACC Press: Washington, DC, 1997.

10. Young, D.S. *Effects of Drugs on Clinical Laboratory Tests*, 5th Ed.; AACC Press: Washington, DC, 2000; Vols. 1&2.

11. Fraser, C.G. *Biological Variation: From Principles to Practice*; AACC Press: Washington, DC, 2001.

12. Lott, J.A.; Smith, D.A.; Mitchell, L.C.; Moeschberger, M.L. Use of medians and "average of normals" of patients' for assessment of long-term analytical stability. Clin. Chem. **1996**, *42*, 888–892.

13. Schultz, E.K. Selection and interpretation of laboratory procedures. In *Tietz Textbook of Clinical Chemistry*, 3rd Ed.; Burtis, C.A., Ashwood, E.R., Eds.; WB Saunders: Philadelphia, 1999; 310–319.

14. Westgard, J.O. *Six Sigma Quality Design and Control*; Westgard QC: Madison, WI, 2001; 285–294.

3

Using Imaging Biomarkers to Demonstrate Efficacy in Clinical Trials: Trends and Challenges

Charles G. Peterfy
Synarc, Inc., San Francisco, California, U.S.A.

Barbara Maley
Eli Lilly and Company, Indianapolis, Indiana, U.S.A.

I. EVOLUTION OF IMAGING BIOMARKERS FOR CLINICAL TRIALS

Interest in imaging biomarkers and their use in clinical trials has literally exploded over the past several years. This is evidenced by the proliferation of symposia and scientific conferences around the world that focus on this topic. The National Institutes of Health and various international societies have formed committees and launched ambitious initiatives to promote the area and advance knowledge about the use of imaging biomarkers in clinical research. Additionally, companies have emerged that specialize in this application of imaging, and a unique service category is forming.

That medical imaging has a contribution to make in drug development is not surprising. Radiology has made remarkable advances over the years, and has been an integral part of day-to-day clinical practice for decades. Indeed, it is hard to imagine any hospital today functioning without a radiology department. Moreover, the potential role that imaging could play in clinical trials is the same role it currently plays in clinical practice, namely patient selection (diagnosis and staging), monitoring disease progression and treatment response, and assessing complications of therapy. What is, perhaps, difficult to understand is why the interest in imaging for clinical trials took so long to develop, as it has only been in the last several years that drug development has benefited substantively from these powerful tools. Why should such a disparity exist between clinical practice and clinical research? Why is radiology so late to this?

At least part of the answer lies in a fundamental catch-22 that the drug development industry must operate under. That is, there is no need in clinical practice for methods of identifying patients that are most appropriate for a particular therapy and for measuring the effectiveness of that therapy before the therapy itself is clinically available. Yet, highly precise and fully validated methods for doing just these things are necessary for developing and gaining regulatory approval of any putative new therapy. Accordingly, the demand for such methodological innovations arises first during the clinical testing of new therapies, and it is therefore, in a Darwinian sense, the priorities and unique regulatory and logistical constraints of the clinical trials process that shape

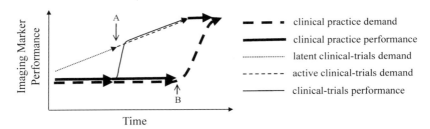

Figure 1   Therapy is a key driver of innovation in medical imaging. The thick broken line (- - - -) depicts the level of performance demanded by mainstream clinical practice for imaging markers of articular cartilage. Prior to the availability of cartilage-modifying therapy, this trajectory is relatively low and flat. The thick solid line (—) shows the actual performance of available imaging techniques and markers (radiographic joint-space width, subjective cartilage integrity with MRI), which tracks the clinical demand closely. The performance required for clinical trials of cartilage-modifying therapy (– – –) is greater, particularly with respect to validity, multicenter stability, and measurement precision, but this is only a latent demand until putative therapies begin to appear in the pipelines of pharmaceutical and biotechnology companies. Entry of these new therapies into clinical testing (A) triggers the development of new imaging techniques and markers (————) along performance criteria aligned with the priorities of clinical trials proven. Initially, the enhanced performance of these imaging endpoints is not valued by mainstream clinical practice. The technology demand in clinical practice, therefore, continues along its original, shallow trajectory. However, once the new therapy enters the market (B), a new demand emerges in clinical practice for methods of selecting patients appropriate for the new therapy and monitoring treatment effectiveness and safety. In this way, the tools and endpoints adapted to suit clinical trials research eventually find their way back into mainstream clinical practice. (Adapted from Ref. 12.)

the early evolution of these radiological techniques (Fig. 1). During this phase of technical/analytical evolution, mainstream clinical practice has little value for the enhanced performance that these methods provide, and it is only once the therapies are approved for clinical use and available to clinicians that a demand for patient-selection and efficacy-monitoring tools emerges in the mainstream. This very dilemma, however, has spawned a new breeding ground for radiological innovation and a novel process for evolving medical imaging technologies.

In osteoporosis research, for example, it was the promise of bisphosphonate therapy that created the initial demand for more precise methods of noninvasively measuring bone mineral density. This demand fueled the evolution of dual-energy x-ray absorptiometry (DXA), a unique imaging technology that prior to the appearance of bisphosphonate therapy had no place in mainstream clinical practice, but that today is part of the routine clinical management of patients with osteoporosis. It was during the clinical trials of these therapies that DXA was refined and validated. Moreover, the availability of DXA and other tools, such as vertebral morphometry, for monitoring disease progression and treatment response in osteoporosis has stimulated further drug development in this area. So, therapeutic innovation and diagnostics are locked in a coevolutionary process, in which advances in one stimulate advances in the other.

A similar pattern is seen in arthritis research, as clinical trials of rheumatoid arthritis employ radiographic scoring methods that are not yet used in clinical practice and that only a few investigators in the world have had any substantial experience with. These semiquantitative scoring methods, such as Sharp scoring, are nevertheless an essential part of the clinical development of any putative disease-modifying therapy for rheumatoid arthritis. Until recently, such therapies did not exist, but over the last couple of years, a number of compounds (e.g., etanercept, leflunomide, infliximab) have been proven in clinical trials using these methods to slow the progression of structural damage in the joints of patients with rheumatoid arthritis. As the clinical availability of these new therapies increases, so will the demand for effective methods of staging and monitoring disease severity and treatment response in patients receiving them.

The same is true in osteoarthritis, in which the recent emergence of structure-modifying therapies in the pipelines of biotechnology and pharmaceutical companies has spawn numerous innovations in imaging evaluation of this disease and its therapy. Tools that have been adapted for multicenter clinical trials are useful not only for global drug development but also for large epidemiological studies and extend more easily into clinical use than do some of the cutting-edge technologies typically used in single-site university research. Accordingly, innovation in clinical-trials radiology can advance understanding about the disease as well. This is exemplified by the recently launched Osteoarthritis Initiative, an historic collaborative between the National Institutes of Health (NIAMS, NIA) and the pharmaceutical industry to conduct a large, longitudinal study aimed at, among other things, at creating new knowledge about the proper use of imaging and other biomarkers in clinical trials of osteoarthritis.

Imaging is also poised to contribute substantively to the development of therapies for neurological disorders, most notably multiple sclerosis, Alzheimer's disease, and stroke, as well as cardiovascular disease and cancer. The following discussion outlines these applications of medical imaging and points to areas where further advances can be anticipated in the near future.

II. THE DIFFERENCE BETWEEN RADIOLOGICAL ASSESSMENT AND CLINICAL ASSESSMENT IN CLINICAL TRIALS

Extracting information from medical images is a complex process. Any medical image offers an abundance information that is immediate to any eye. However, that information is useless until it can be translated unambiguously into words or numbers (Fig. 2). The process of extracting clinical information from medical images is referred to in clinical radiology as "image reading." Reading is an entirely appropriate name for this process, as it is not unlike the process involved in reading text. In both cases, one must learn to recognize a set of symbols in

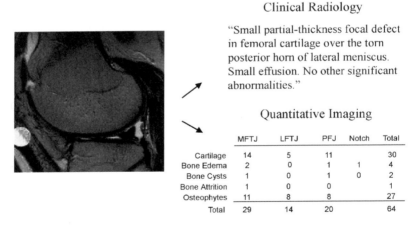

Clinical Radiology

"Small partial-thickness focal defect in femoral cartilage over the torn posterior horn of lateral meniscus. Small effusion. No other significant abnormalities."

Quantitative Imaging

	MFTJ	LFTJ	PFJ	Notch	Total
Cartilage	14	5	11		30
Bone Edema	2	0	1	1	4
Bone Cysts	1	0	1	0	2
Bone Attrition	1	0	0		1
Osteophytes	11	8	8		27
Total	29	14	20		64

Figure 2   Translating images to numbers. Extracting relevant information from medical images involves a process very similar to reading. Clinical radiologists become image "literate" only after years of training and practice. Translating image data into numerical form, as is needed for clinical research, requires additional expertise and training, and often the use of specialized computer algorithms that cannot be found in mainstream clinical practice.

a vast array of patterns and understand the different patterns in a context of meaning. It is easy to forget how much time and effort was required to make this process automatic in text reading, and the importance that literacy has in our day-to-day lives. Medical-image reading is no different. Residency training in radiology in the United States takes 4 years, and at the end of this training a certified radiologist can extract the pertinent clinical information from a magnetic resonance (MR) image in a matter of only moments. To the image-illiterate observer, the process can seem mysterious—even mystical—but to the clinicians and patients who rely on accurate radiological reading, the value of image literacy is undeniable.

Image literacy is thus an important requirement for clinical-trials radiology. But, it is still just a prerequisite. Clinical radiologists are expert in extracting clinically relevant information from medical images and expressing them in words. Additional training, expertise, and, occasionally, specialized software are required to extract the morphological, compositional, and process-related information relevant to clinical trials, and further yet to express this information unambiguously in numerical form. Unfortunately, radiologists with the requisite expertise and experience are extremely rare, and clinical-trials radiology will probably remain a superspecialized niche of the mainstream specialty for some time to come—but one with a substantial contribution to make to drug development and the advancement of medicine, nevertheless.

One of the key challenges facing clinical trials and epidemiological research is dealing with multi-center data collection. This represents a fundamental departure from traditional university research, which typically operates on a single-site basis, but it is a necessity when hundreds or thousands of patients must be evaluated rapidly and efficiently in a single study. Radiology offers a unique advantage over clinical assessments in this regard, as it allows centralization of data generation, not just data management (Fig. 3).

In a typical multicenter clinical trial hundreds or potentially thousands of patients must be evaluated at multiple sites throughout the country or around the world in a consistent and timely fashion. At each of these sites, clinical investigators, using their medical expertise and judgment, extract the relevant clinical information from their subset of patients and record the results on case-report forms that are then aggregated and databased by a central facility, such as a contract research organization (CRO). The multiplicity of sources of data generation (clinical investigators) in this scenario is an inescapable source of variability in clinical assessment data. Centralizing the data management is essential to any large multicenter study, but does not solve this fundamental problem.

Radiological assessment, however, is able to centralize the actual data generation and thereby contain this source of variability. In this scheme, the expert central radiologist, based on the scientific and regulatory needs of

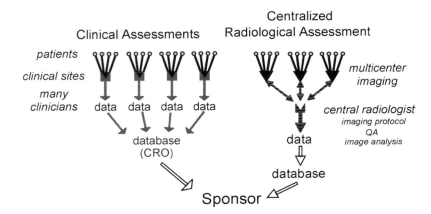

Figure 3 Clinical-trials radiology centralizes data generation, not just data management. The multiplicity of sources of data generation (clinical investigators) in the clinical assessments of multi-center trials is an inescapable source of variability. Centralized data management is essential to any large multicenter study, but does not solve this problem. Radiological assessment, however, is able to centralize data generation and thereby contain this source of variability. In this scheme, the expert central radiologist, based on the scientific and regulatory needs of the study, the capabilities of the sites selected, and an intimate understanding of how he/she will analyze the images, designs an imaging protocol that will generate the type of images needed consistently across all the patients, sites, and equipment platforms, and throughout the duration of the study. The radiologist also provides any information or training necessary to the sites on an ongoing basis, and then checks the images to ensure protocol compliance and adequate image quality. Finally, using special expertise and sophisticated computer aids, the central radiologist extracts the relevant morphological, compositional and physiological information from the images and enters it directly in to the central database. Centralizing data generation in this way not only reduces variability in radiological assessments but facilitates integrating the critical components of this process, as image acquisition, quality assurance (QA) and image analysis/quantification must all come together seamlessly to work properly.

the study, the capabilities of the sites selected, and an intimate understanding of how he/she will analyze the images, designs an imaging protocol that will generate the type and quality of images needed, consistently across all the patients, sites, and equipment platforms in the study, and for the entire duration of the study. This may include the use of specialized image-acquisition aids (IAA) designed to improve the consistency of multicenter imaging (see below). The central radiologist also provides any necessary information or training to the sites on an ongoing basis, and checks the images to ensure protocol compliance and adequate image quality. Finally, using special expertise and sophisticated computer algorithms, the central radiologist extracts the relevant morphological, compositional, and process information from the images and enters the results directly into a central database. Centralizing data generation in this manner not only reduces variability in radiological assessments but facilitates integrating the essential components of this process.

In clinical-trials radiology, as in conventional clinical radiology, image acquisition, image analysis, and quality assurance must all come together seamlessly into a single integrated process to work properly. It is difficult to dissociate these elements, even in routine clinical practice, but in clinical-trials radiology proper integration is essential. Moreover, this integration must be right to left. That is, it is the image analysis method that dictates how the images must be acquired and what quality-control considerations must be focused on. The imaging must furthermore be integrated with the clinical assessments and any molecular marker measurements or other tests included in the study. Integrating all of these elements properly in a clinical trial requires special expertise, experience, and systems.

The priorities of clinical-trials research are also slightly different than those of clinical practice. In clinical practice, the principal objective is to diagnose accurately and optimally treat a specific patient. In clinical trials, the patient is an abstraction, defined wholly as a set of numbers in a database, and the objective is to demonstrate formally the efficacy, safety, and cost utility of a new therapy for regulatory approval, with the fewest patients and in the shortest time possible. These different contexts embody slightly different selective pressures and performance metrics for imaging techniques and markers, and therefore shape the character of these markers differently. As stated earlier, it is important to understand these differences to be able to anticipate the attributes of the techniques and markers that will work best in each of these environments (Fig. 1). The following sections outline the key considerations that go into selecting the right imaging techniques for a particular clinical trial, and how to evaluate the relative utility of one method over another in a particular research or clinical context.

III. CLASSES OF IMAGING MARKERS

Aside from their statistical classification as nominal, ordinal, or continuous data, imaging markers can be categorized as morphological, compositional, or process related. Morphological markers relate most closely to the traditional concept of anatomical imaging and include a variety of dimensional and geometrical measures. Examples include radiographic vertebral morphometry in osteoporosis; radiographic OARSI (Osteoarthritis Research Society, International) scoring [1] or whole organ MRI scoring (WORMS) [2,3] in osteoarthritis, radiographic Sharp scoring of bone erosions and joint-space narrowing in rheumatoid arthritis [4,5]; sonographic measurement of carotid intima-media thickness; MRI measurements of entorhinal cortex and hippocampal volume in Alzheimer's disease [6]; and dimensional measurements of tumor size in cancer. The technical challenge in all of these morphological assessments is edge detection and image segmentation. These depend on spatial resolution and image contrast between the structure of interest and background tissue, and both of these image parameters in turn depend on the details of the acquisition technique used. Therefore, success in morphological image analysis requires sophistication and experience in designing imaging protocols as well as a clear understanding of how different protocol decisions will affect the measurement algorithm to be used (Figs. 4 and 5).

A subcategory of morphological markers are measures that relate to the microstructural integrity of tissues. An emerging MRI technique, known as diffusion-weighted imaging, falls into this category. Diffusion-weighted imaging derives from the physical, random (Brownian) motion of water molecules. Importantly, the microenvironment of water molecules critically influences the freedom, or "mean free path," of diffusion. Intra- and extracellular apparent diffusion coefficients differ markedly. The "effective" diffusion coefficient, measured across an image pixel, typically represents an average of the contained diffusion environments, allowing delineation of regions of cellular swelling (e.g., in ischemia) or necrosis. The close correlation of diffusion-weighted images and derived apparent diffusion coefficient maps of tumors to postmortem histological sectioning and staining suggests a role for diffusion weighted imaging as a tool for virtual biopsy in vivo (Fig. 6). Diffusion-weighted imaging is also useful in stroke, as early acute changes in stroke associated with cytotoxic edema affect the microenvironment of local water molecules. Measuring the volume of regions of hyperintense signal in diffusion-weighted MR images captures this change and thereby helps estimate ischemic lesion size. A refinement of diffusion-weighted imaging is diffusion-tensor imaging. This technique maps the direction of water diffusion within anisotropic tissues, such as the brain (Fig. 7). Decreased anisotropy in the brain indicates disruption of white matter fiber tracts in neurodegenerative disorders, such as multiple sclerosis and Alzheimer's disease. Micro-CT and micro-MRI measurements of trabecular density, trabecular connectivity, and fractal dimension in bone are examples of microstructural markers applicable to osteoporosis [7,8]. In arthritis imaging, increased T2 relaxation of articular cartilage on MRI reflects disruption of the micro-organization of matrix collagen fibrils [9,10] (Fig. 8). Novel optical imaging techniques, such as optical coherence tomography (Fig. 9), not only provide very high spatial resolution (10 μm) but can also be combined with absorption or polarization spectroscopy to probe the microstructural integrity of thin tissues, such as articular cartilage or vascular walls.

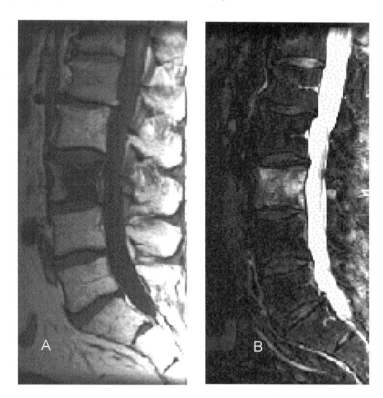

Figure 4 Image acquisition dictates the scope of image analysis possible. Delineating the margins of a lesion for dimensional measurements depends on image contrast and spatial resolution. (A) T1-weighted MRI image of a metastatic lesion in L3. The lesion margins are well-defined because of the intrinsic contrast of the low signal intensity of the lesion against the high signal intensity of the residual marrow fat. The fat-suppressed T2-weighted image of the same spine (B) shows greater contrast and a correspondingly greater extent of involvement of the vertebral body by tumor. Accordingly, decisions about the image acquisition technique affect the accuracy and precision of dimensional measurements. (Courtesy of Synarc, Inc., with permission.)

Other imaging markers provide information about the biochemical composition of tissues. In contrast to biochemical markers assayed in serum, urine, or other body fluids, compositional imaging markers map the spatial distribution of tissue constituents, and can in some cases quantify steady-state tissue concentrations of certain constituents. Bone mineral density measured with DXA is an example of this class of marker used in the study of osteoporosis. In neurodegenerative diseases, such as Alzheimer's disease and multiple sclerosis,

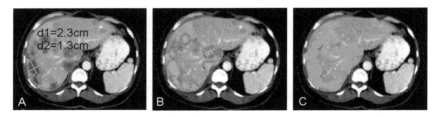

Figure 5 Computer-assisted image analysis. (A) CT image of a liver containing several metastatic lesions and the orthogonal dimensions (cross-product $= 3.0\,cm^2$) of one lesion in the right lobe. (B) Computerized segmentation of the lesions to measure their volumes. However, because of improper thresholding several lesions are not detected and the volumes of the ones that are identified are underestimated. (C) Reader-corrected segmentation allowing accurate volume quantification. Optimal performance in image analysis thus comes from integrating expert judgment from the clinical-trials radiologist with the computing power of the measurement algorithm. (Courtesy of Synarc, Inc., with permission.)

MR spectroscopy and spectroscopic imaging of n-acetylaspartate, an amino acid specific in the brain to neuronal tissue, combined with spectroscopic measurements of myo-inositol, a glial marker, can greatly increase the sensitivity and specificity of brain atrophy measurements (Fig. 10). Coronary calcification score using electron-beam computed tomography (CT) or spiral multislice (multirow detector) CT is a compositional marker shown to be an earlier and

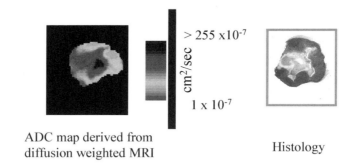

ADC map derived from
diffusion weighted MRI

Histology

Figure 6 Diffusion-weighted MRI in an experimental RIF-1 tumor in a rodent model. Diffusion-weighted MRI allows delineation of the necrotic center (higher apparent diffusion coefficient [ADC]) compared to surrounding viable tissue (lower ADC). Correlation with histological sectioning suggests the role of diffusion-weighted MRI as a "noninvasive biopsy." (Courtesy of K. Helmer, Worcester Polytechnic Institute.) (See color insert.)

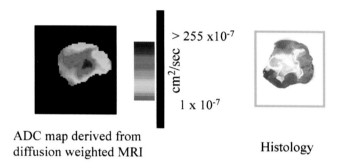

ADC map derived from
diffusion weighted MRI

Histology

Figure 3.6 Diffusion-weighted MRI in an experimental RIF-1 tumor in a rodent model. Diffusion-weighted MRI allows delineation of the necrotic center (higher apparent diffusion coefficient [ADC]) compared to surrounding viable tissue (lower ADC). Correlation with histological sectioning suggests the role of diffusion-weighted MRI as a "noninvasive biopsy." (Courtesy of K. Helmer, Worcester Polytechnic Institute.)

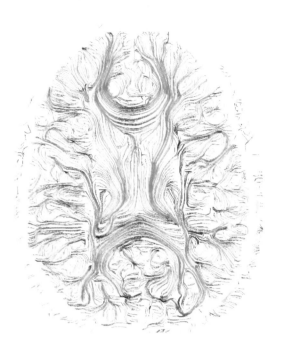

Figure 3.7 Diffusion-tensor MRI. Vector map showing the preferred orientation of white-matter tracts in the human brain from a diffusion-tensor imaging examination. The color map shows the orientation of the tracts in a medial-lateral (red), anterior–posterior (green), and through plane (blue) composite. (Courtesy of M. Moseley, Synarc, Inc. and Sanford University.)

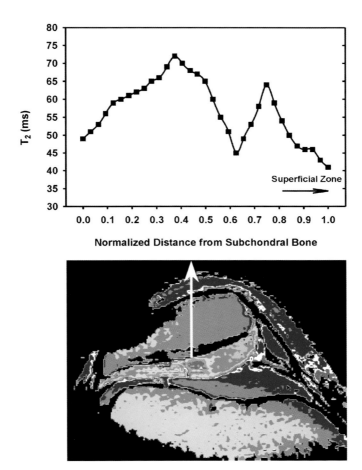

Figure 3.8 Cartilage-T2 mapping with MRI. (Lower panel) Axial T2 relaxation map of the articular cartilage of the patella generated with multiecho MRI at 3T. A focus of decreased T2 relaxation near the ridge is indicative of collagen matrix damage. The graph above shows the T2 profile of a line through this region of articular cartilage. (Courtesy of B. J. Dardzinski, Ph.D., University of Cincinnati College of Medicine.)

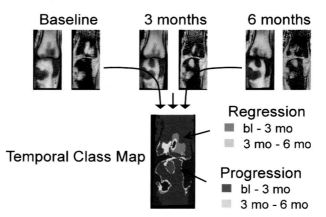

Figure 28 Multispectral and temporal data fusion. Fusion images generated from spatially registered MRI acquired with different pulse sequences and at different times display tissue classes and changes in different colors. This may facilitate image interpretation in longitudinal studies and provide an audit trail for regulatory purposes. (From Ref.12.)

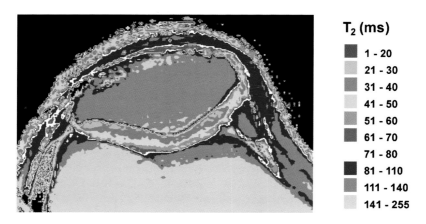

Figure 29 T2 relaxation map of normal adult articular cartilage. T2 map generated from multislice, multiecho (11 echoes: TE = 9,18,... 99 ms) spin echo images acquired at 3 T shows increasing T2 toward the articular surface. (Courtesy of B. J. Dardzinski, Ph.D., University of Cincinnati College of Medicine.)

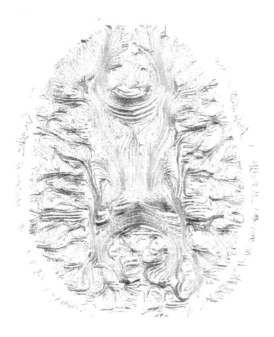

Figure 7 Diffusion-tensor MRI. Vector map showing the preferred orientation of white-matter tracts in the human brain from a diffusion-tensor imaging examination. The color map shows the orientation of the tracts in a medial-lateral (red), anterior–posterior (green), and through plane (blue) composite. (Courtesy of M. Moseley, Synarc, Inc. and Sanford University.) (See color insert.)

more precise predictor of cardiac events in asymptomatic patients with atherosclerotic risk factors than any other noninvasive screening method, including stress electrocardiogram (EKG), stress echocardiography, and thallium scintigraphy [11]. T2 relaxation is a compositional MRI marker of collagen content in fibrous tumors, such as fibrosarcoma, desmoid tumor, and neurosarcoma. As mentioned earlier, T2 relaxation can also be used as a measure of collagen organization and content in articular cartilage [10,12]. Proteoglycan content in articular cartilage also can be quantified by MRI in terms of the fixed negative charge density of the glycosaminoglycan moieties. This can be done by quantifying sodium concentration in cartilage using sodium MRI, as sodium is the primary cation in cartilage balancing the negative charge of constituent proteoglycans [13]. Alternatively, proteoglycans can be quantified by measuring the concentration of negatively charged MRI contrast agent, $Gd\text{-}DTPA^{2-}$ (through its effect on T1 relaxation) imbibed by cartilage in inverse

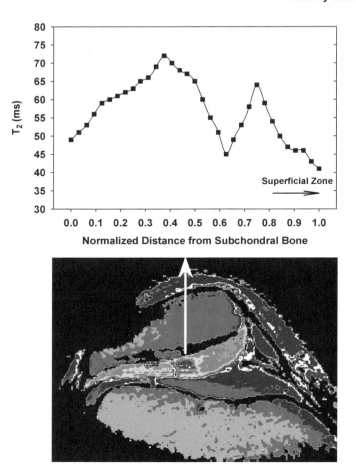

Figure 8 Cartilage-T2 mapping with MRI. (Lower panel) Axial T2 relaxation map of the articular cartilage of the patella generated with multiecho MRI at 3T. A focus of decreased T2 relaxation near the ridge is indicative of collagen matrix damage. The graph above shows the T2 profile of a line through this region of articular cartilage. (Courtesy of B. J. Dardzinski, Ph.D., University of Cincinnati College of Medicine.) (See color insert.)

proportion to the fixed negative charge density of the tissue [14,15] (Fig. 11). Cationic contrast agents have also been shown to shorten articular cartilage T1 in proportion to proteoglycan content [16].

Process-related imaging markers include measures of tissue perfusion, blood volume, and microvascular permeability. These microvascular markers are promising tools in cancer clinical trials, particularly for angiostatic therapies that

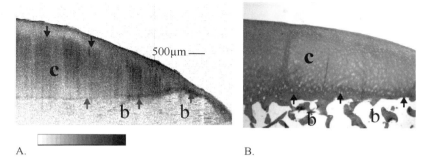

Figure 9   Optical coherence tomography of articular cartilage. Optical coherence image (A) and corresponding histological section (B) demonstrates the exquisite spatial resolution of this technique. (Courtesy of Mark Brezinski, with permission.)

halt tumor growth without necessarily reducing lesion size. Markers of perfusion include ⁹⁹ᵐTc-HMPAO SPECT, various parameters derived with Doppler ultrasound, and changes in T1 to T2* relaxation on dynamic MRI following bolus intravenous injection of Gd-DTPA. Perfusion deficits in stroke patients define

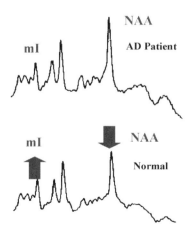

Figure 10   MR spectroscopy in Alzhemimer's disease. Proton MR spectroscopy is a powerful tool for evaluating metabolities in vivo. Measurements are usually made on a single voxel position in the region of interest. The ratio of *n*-acetylaspartate (NAA) to *myo*-inositol (mI) concentration (NAA/mI) has been shown to correlate with cognitive abilities in patients with probable Alzheimer's disease and in age-matched controls. (Courtesy of Synarc, Inc. with permission.)

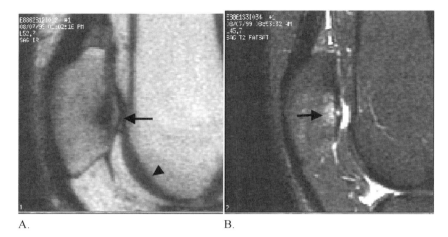

A.   B.

Figure 11   MRI markers of cartilage matrix integrity. (A) Sagittal inversion-recovery image of a knee following intravenous administration of Gd-DTPA shows a region of high signal intensity (arrow) in the patellar cartilage indicative of abnormal uptake of anionic Gd-DTPA, and therefore, local proteoglycan depletion. Cartilage in the trochlear groove (arrowhead) shows low signal intensity indicative of repulsion of Gd-DTPA by negatively charged proteoglycans. (B) Fat-suppressed, T2-weighted image of the same knee prior to Gd-DTPA injection shows a smaller focus of increased signal intensity (arrow) in the same location indicative of local collagen matrix loss. This is associated with subarticular marrow edema in the patella. (From Peterfy CG. The role of MR imaging in clinical research studies. Semin Musculoskelet Radiol 5(4): 365–378, 2001.)

the region of tissue at risk for infarction and offer a potential endpoint for thrombolytic therapy. When combined with diffusion-weighted imaging, the perfusion/diffusion mismatch provides a marker for salvageable tissue. Alzheimer's patients exhibit regional perfusion deficits that have been shown to be predictive of future decline in cognitive ability, providing a method for enriching study populations with rapid progressors. Unfortunately, MRI-derived blood volume and vascular permeability measurements are currently restricted to the brain, where the blood-brain barrier normally restricts extravasation of Gd-DTPA. When this barrier is disrupted by inflammation or tumor neovascularity, Gd-DTPA diffuses into the local interstitium (Fig. 12). Because of its small molecular size, Gd-DTPA readily diffuses out of even normal vessels, precluding accurate estimation of blood volume (a measure of vessel density), permeability surface-area product, or fractional leak rate. These parameters can, however, be determined using macromolecular contrast agents, such as polylysine-chelated or albumin-chelated Gd-DTPA [17,18] (Fig. 13). Although, these macromolecular contrast agents are not currently approved for use in

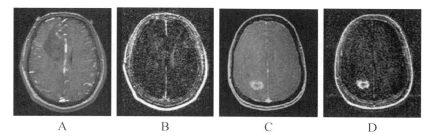

A	B	C	D

Figure 12 Permeability mapping of brain tumors with MRI. Source MRI (A) and synthesized permeability map (B) from a low-grade brain tumor (grade II astrocytoma) show no significant difference in permeability between the tumor and healthy tissue. Soruce MRI (C) and permeability map (D) from a grade IV tumor (glioblastoma muliforme) *quantitatively* reveals a rine of high microvascular permeability, indicative of angiogenesis. (Courtesy of Heidi Roberts, with permission.)

humans, a number of companies are actively developing similar contrast agents that could be used for microvascular assessments. These techniques can also be used to evaluate synovium [19] and pre-erosive osteitis in rheumatoid arthritis and ischemic changes in the heart.

Nuclear medicine techniques such as positron emission tomogrpahy (PET) provide exquisite sensitivity for biochemical processes, such as metabolism. Using a radioactive tracer, such as fluorodeoxyglucose (^{18}FDG), the metabolic pathway of glycolysis can be followed and focal regions of tracer accumulation

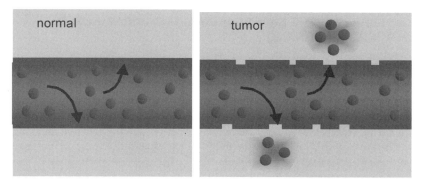

Figure 13 Identifying neovascularity with macromolecular MRI contrast. Conventional MRI contrast agents leak rapidly from both normal and abnormal extracranial vessels. However, macromolecular contrast media permeate only abnormal vascular walls in areas of inflammation or neovascularity.

can be monitored. An extensive array of novel radiolabeled probes is under development allowing numerous biochemical pathways and processes to be monitored and localized. Hypo- or hypermetabolism can be quantified and related, via multimodality image fusion, to results from other imaging modalities building up a composite picture of tissue dysfunction at a structural, cellular, vascular, and metabolic functional level.

Much of the development of these types of process markers is taking place in the emerging field of molecular imaging. Molecular imaging differs from conventional techniques in that it identifies specific gene products and intracellular processes using picomolar or micromolar quantities of specialized imaging probes. Reporter gene imaging exemplifies this approach and targets cell surface proteins or receptors, or intracellular enzyme activity, such as p53 tumor suppressor gene expression, initiated by therapy. A particularly intriguing aspect of molecular imaging research is the design of activated MR imaging agents. These imaging agents are engineered to remain inactive until "turned on" by specific enzymes, such as caspase or matrix metalloproteinases, to provide early detection of the onset of apoptosis, a cell cluster's transformation into cancer, or some other critical pathophysiological or therapeutic process of interest. Accordingly, molecular imaging shows great promise for clinical trials, provided researchers can figure out how to get these large molecules into cells and how to increase their MR signal potency sufficiently to image small concentrations of probe.

Aside from the theoretical implications of some of these innovations, a number of practical factors must be considered in selecting an imaging marker for a specific purpose in a clinical trial. The following section addresses this.

IV.   SELECTING THE RIGHT IMAGING TECHNIQUE FOR THE PURPOSE AT HAND

Selecting imaging endpoints for a clinical trial requires consideration of the scientific and regulatory needs of the study, the specific competencies and imaging equipment available at the clinical sites involved, the image processing and analysis methods that will be used ultimately to generate the study data, and of course, the budget. Two key design considerations are the role that the imaging will play in the trial and the role that the trial itself will play in the overall development program of the compound.

As in conventional clinical radiology, imaging can play three fundamental roles in a clinical trial: (1) patient selection, (2) monitoring treatment efficacy, and (3) monitoring treatment safety. The performance specifications to which an imaging marker, or for that matter any biomarker, must be held depends on which of these roles it is asked to play. Both diagnostic and prognostic biomarkers are used for patient selection. The latter help predict which patients are most likely to progress most rapidly, respond to a particular therapy, or develop complications from the therapy. In clinical practice identifying rapid progressors helps determine which patients are most in need of aggressive therapy. In clinical trials, enriching the study population with rapid progressors can reduce the time required to demonstrate a difference in therapeutic effect. This can be particularly useful in proof-of-concept and dose-selecting studies, in which rapid readout of the results is usually a priority. It is also useful when the toxicity of prolonged use of an investigative compound is still uncertain. Increasing study group homogeneity through patient selection increases statistical power and thus reduces the number of patients needed to detect differences between treatment and comparitor arms. This can be important when the test compound is in limited supply or very costly to produce. On the other hand, rapidly progressing patients may be more recalcitrant to therapy or yield results that are less generalizable to the desired treatment population. The immediate and long-term practical and regulatory implications of a patient selection strategy must, therefore, be given careful consideration during the design of a drug development program.

Patient-selection markers fall into three basic categories: (1) those that identify a disease process (e.g., MRI markers of synovitis), (2) those that identity specific tissue or patient characteristics (e.g., tumor estrogen receptors), and (3) those that establish the severity and extent of disease, or the disease burden (e.g., bone mineral density, tumor stage, Sharp score). These markers are usually also useful for monitoring treatment response, but not always. The critical criterion for a marker to be useful for monitoring treatment is that the marker change with the disease and/or therapy. This is not necessarily the case with markers that identify specific tissue or patient characteristics, or for markers of disease features upstream from the site of action of the therapy being tested (Fig. 14). In addition to whether the maker actually changes with disease and/or therapy, the rate at which it changes and how precisely this change can be measured are important considerations. Perhaps the most important factor, however, is the link between measurable changes in the putative marker and the clinical outcome of interest (see below).

Finally, imaging can be used to evaluate the safety of a putative therapy. In this case, it is important to distinguish complications of the therapy from those associated with the underlying disease. Prognostic marker that are able to predict which patients will develop complications are extremely valuable, as they can be used to reduce the rate of adverse effects in clinical practice postapproval. In some cases, the availability of such a marker can mean the difference between regulatory approval and program failure.

Another important consideration in selecting an imaging marker is the role that the study itself is intended to play in the overall drug development program. These roles include: (1) internal decision making for drug portfolio management

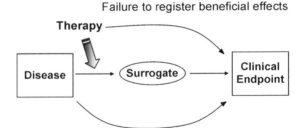

Figure 14 Surrogate validity. Use of a (bio)marker as a surrogate endpoint assumes that the designated biomarker lies directly along the disease pathway to the true outcome of interest. To be useful in therapeutic trials, the surrogate must also lie on the intervention pathway. Alternative mechanisms of disease or therapy that bypass this biomarker undermine its validity as a surrogate endpoint.

and future-study design, (2) definitive demonstration of efficacy and safety for regulatory approval, (3) educating clinicians and patients about the therapeutic value and proper use of the compound, and (4) mechanistic and epidemiological investigation aimed at creating new knowledge about the disease and its therapy.

Well-established imaging markers that are already accepted by regulatory agencies for definitive testing in phase III, e.g., vertebral morphometry for osteoporosis, tumor diameter for cancer, or Sharp scoring of the hands for rheumatoid arthritis, can also be used for internal decision making. However, more innovative but less fully validated markers, such as bone mineral density, tumor vascular permeability, or synovial volume, may offer substantial advantages in terms of speed or statistical power over these more conservative endpoints. Effective use of imaging markers for proof-of-concept in phase I or II can prevent costly mistakes by helping eliminate compounds that lack sufficient efficacy before committing to phase III testing. These markers can also help optimize dose and dosing regimen, sampling intervals, and patient types for confirmatory phase III studies. Proper planning at this stage can greatly reduce uncertainty about entering phase III, decrease the time to registration, and help contain overall costs of drug development. Therapeutic confirmatory studies are tightly focused on gaining regulatory approval and rely more heavily on fully validated and broadly accepted endpoints. Therefore, success in a drug development program depends not only on effective project management, data management, and image analysis techniques, but proper planning in exploratory phase Ib and II studies.

Another important role of many trials in a drug development program is to educate clinicians, health management organizations, and patients about the therapeutic value and proper use of the compound once regulatory approval is granted. This includes supportive evidence of efficacy and safety and instruction on therapeutic use in a diversity of clinical settings and circumstances. Endpoints that are the most useful for interval decision making or convincing for regulatory approval are not necessarily the most illustrative or understandable to clinicians and patients who will ultimately use the product. Imaging is particularly useful in this regard and can be a powerful tool for marketing a new therapy. Additionally, clinicians must be able to identify patients most appropriate for the therapy and to monitor individual patients' responses to treatment. It is important, therefore, that the markers and techniques used in clinical trials be integrated effectively back into clinical practice following approval. Proper planning for this at the outset of a development program can help expand distribution and increase peak sales.

Finally, since the rigor and statistical power employed in clinical-trials imaging are typically far greater than those used in traditional university studies, clinical-trials imaging data also offers a rich substrate for further discovery about a disease and its therapy. Large relational databases can provide an invaluable resource for rational protocol development, site selection, and choice of patient entry criteria.

A. General Performance Metrics for Imaging Markers

Based on the considerations outlined above, a fundamental set of performance criteria can be defined by which the utility of different imaging markers can be compared in a variety of clinical and research applications. These metrics are: (1) validity, (2) responsiveness (rate of change) to disease and therapy, (3) measurement precision, (4) convenience, and (5) cost.

As stated previously, the most important measure of marker's utility in clinical trials or clinical practice is its link to the "true" clinical outcome of interest. Very few biomarkers thus far have been widely accepted as true surrogate endpoints for clinical outcomes. However, the degree of validity required depends on the specific objectives of the study. Therapeutic confirmatory studies, upon which regulatory approval will be based, demand the most stringent validation and regulatory acceptance of the primary endpoints used. Secondary endpoints or those used in studies aimed primarily at internal decision making can take advantage of more novel techniques and markers that may not yet be formally ratified but that carry significant advantages in terms of predictive power and sensitivity to change.

In addition to the pathophysiological validity of the disease feature itself, it is important to consider the technical, or assay, validity of the instrument used to measure it. This relates to the image acquisition technique and the reading/measurement system used, and is typically expressed in terms of sensitivity, specificity and area under the receiver operating characteristic (ROC) curve, provided a gold standard exists. In the absence of hard criterion validity, softer, indirect validation measures, such as expert opinion or group consensus, must be relied on.

It is also important that the dynamic range of the surrogate marker capture as much of the clinically relevant range of changes in the "true" outcome as possible (Fig. 15). Insensitivity of a surrogate marker to early changes in the true outcome is sometimes called a "floor effect" while failure to register severe but still relevant changes in the true outcome is known as a "ceiling effect." Radiographic minimum joint-space width, which is an indirect marker of articular cartilage thickness in diarthroidial joints, shows a ceiling effect when additional cartilage loss is still possible in a joint compartment after the space has been completely obliterated in one particular spot. Dynamic range characteristics can influence study results in complex ways. Consider the following scenario (Fig. 16). Hypothetical techniques X and Y both measure the same disease feature (e.g., bone erosion). Technique X shows greater sensitivity than technique Y for small changes but equivalent sensitivity for large changes. Paradoxically, the floor effect of technique Y could make it appear to be more responsive to change than technique X under certain circumstances. It is important, therefore, to consider dynamic range in the design and interpretation of studies containing longitudinal image data.

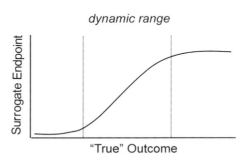

Figure 15   Dynamic range. Dynamic range refers to the proportion of change in the "true" outcome of interest that is captured by changes in the surrogate endpoint. Inability of the surrogate endpoint to detect small changes in the true outcome is sometimes referred to as a "floor effect." Failure to register severe changes is called a "ceiling effect." Ideally, any such floors or ceilings lie outside of the range of changes in the true outcome that are relevant to the question under study.

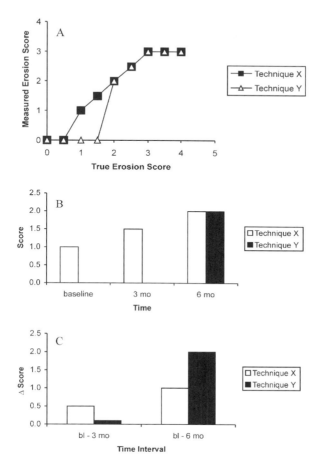

Figure 16 Dynamic-range effects. (A) Technical dynamic ranges of two hypothetical techniques for imaging bone erosions. Technique X is insensitive to erosions smaller than Score 1 (hypothetical scoring system) and technique Y is insensitive to erosions smaller than Score 1 (hypothetical scoring system) and technique Y is insensitive to erosions smaller than Score 2, but both show equivalent performance for erosions greater than or equal to Score 2 (i.e., the techniques show nonlinear sensitivity profiles). If a patient had a Score 1 erosion at baseline, as in (B), it would be registered by technique X but not technique Y. If the erosion grew to grade 2 by 6 months, both techniques would register the erosion at this time point. However, in a plot of the change in erosion score over baseline (C), technique Y, by virtue of its floor effect, would paradoxically register greater progression between baseline and 6 months than would the more sensitive technique X. This hypothetical example illustrates the importance of considering dynamic range in the design and interpretation of longitudinal imaging studies.

How quickly the marker changes in response to disease or therapy is another important metric of performance. Marker responsiveness determines the minimum follow-up interval theoretically possible for demonstrating disease progression or improvement. Highly responsive markers are important in clinical practice for identifying patients who are failing therapy and may require dose adjustments or change to a different, hopefully more effective treatment. Marker responsiveness is also important during the early clinical testing of a new drug when the safety profile has not yet been fully established. Typically, such trials need to be less than 3 months in duration, and therefore require markers that can demonstrate change within that time frame. Additionally, there is enormous financial incentive to accelerate the drug development process and enter the marketplace sooner. This includes first-mover advantages for novel agents, but also longer market exclusivity during the finite life span of a drug patent. The patent for a new investigational drug must be filed as soon as human testing of the drug begins, and therefore a portion of the period of market exclusivity provided by the patent will be consumed by formal clinical testing and regulatory due diligence. When measured in terms of lost revenues, this can amount to $25 million for every month that a $300 million/year drug is delayed entry into the market.

In addition to the rate at which a market changes, how precisely that change can be resolved is an important parameter. Measurement precision thus determines the magnitude of change that can be resolved with confidence, and therefore the marker's sensitivity to change (change-to-error ratio). Sources of precision error include interindividual variation, variability of the method used to acquire the images, and errors stemming from the actual measurement method used. Measurement precision in large multicenter trials is maximized by careful patient selection and use of homogeneous study populations. It is also important to use image acquisition methods that are widely available, stable over time and different equipment platforms, easy to perform, and well tolerated by patients.

In addition to expertise and experience in designing and implementing specialized imaging protocols for multicenter trials, specialized IAAs can help improve image quality and consistency in clinical trials. These IAAs include positioning devices designed to optimize patient positioning, minimize patient movement, and maximize reproducibility of serial examinations (Fig. 17). Other IAAs, known as "phantoms," are external standard references used to test and correct the stability of imaging equipment through out the duration of a trial, cross calibrate different equipment platforms used in the trial and occasionally to correct errors or limitations in the raw image data acquired at different sites (Fig. 18).

Finally, measurement precision error is minimized by centralized data management and image analysis using maximally controlled conditions and highly trained readers coupled with the most powerful image-processing and analysis tools. As discussed earlier, centralized reading can support more

Figure 17 Image acquisition aid for radiography of the knee in clinical trials of osteoarthritis. This Plexiglas frame (SynaFlex, Synarc) was designed to position the feet and knees properly and reproducibly for fixed-flexion radiography of the knees in clinical trials. Down the center of the frame is a phantom (arrows) that allows verification of the x-ray beam angle used and quantification of changes in magnification factor that can result from changes in equipment or human error. (Courtesy of Synarc, Inc. with permission.)

complex and demanding scoring methods and quantitative analyses than would be feasible in clinical practice. Clinical practice typically demands rapid turnaround and therefore on-site readings or efficient teleradiology services. Readings for clinical trials, in contrast, are usually not needed until all of the patients have completed the study, and therefore readings can be done in batches by a remote central facility. Increased measurement precision can be traded for decreased study duration and the number of patients and sites required to test the hypothesis. In addition to the financial upside of early market entry for commercial products, reducing a clinical trial by 200 patients can save more than $1–2 million in direct costs.

Both the responsiveness of a marker and the precision error associated with measuring its rate of change affect longitudinal sensitivity. For a given technique, the smallest change detectable with 95% confidence is 2.8 times the precision

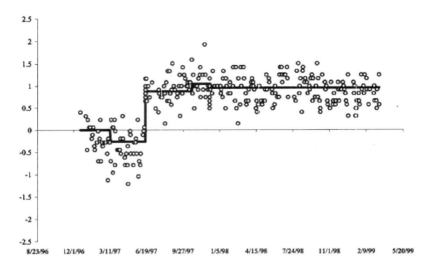

Figure 18   Importance of calibration phantoms in DXA. Longitudinal quality-control data from a DXA scanner showing an abrupt shift in calibration associated with failure of a system component. With use of a standardized bone mineral density test object or "phantom" the calibration of each densitometer in a clinical trial can be measured each day a subject is scanned or at a minimum of three times per week. By comparing these daily measurements against control limits, the DXA technologist can identify scanner problems and request service from the DXA manufacturer. By collecting, reviewing, and analyzing these data, the central radiology service can further evaluate scanner calibration history.

error (e.g., root-mean-square standard deviation for replicate measurements)[20]. To reach 80% confidence (two-tailed, or 90% one-tailed confidence), a change of only 1.8 times the precision error is needed [21]. This less stringent change criterion has been referred to as the trend assessment margin. The time interval required to reach either threshold is determined by dividing the change criterion by the responsiveness of marker (median rate of change per year). The shorter this follow-up interval, the greater the longitudinal sensitivity of the marker. Differences in longitudinal sensitivity among different markers can also be expressed in terms of precision error if corrected for difference in responsiveness. Thus, the standardized precision error for technique A can be expressed as its raw precision error divided by the response ratio of technique A relative to technique B (response rate A/response rate B) [21]. Response ratios are less cohort dependent than response rates because part of the cohort bias cancels out.

Accordingly, a marker that shows twice the precision error but four times responsiveness will still show half the standardized precision error.

Convenience and cost are factors that track with availability, examination time, patient tolerance, and ease of image data transfer, storage, and processing. It is important to optimize the cost-benefit ratio of each element of a study in the context of the overall development program, as some techniques with higher unit costs may actually help contain cost in other areas and/or yield greater benefit— not to mention save time, effort, and frustration.

V. OPTIMIZING MULTICENTER IMAGE ACQUISITION

Image analysis, regardless of the sophistication and talent of the readers or the power of the image-processing and analysis software used, can only be as good as the quality of the original images. Use of improper imaging technique or the presence of serious artifacts can render image data useless. Good image analysis therefore begins with good image acquisition and careful quality control. Performing this properly on a multicenter or multinational basis can be extremely challenging, and requires special expertise and systems that cannot be found in mainstream clinical practice or conventional contract research organizations.

Multicenter clinical trials imaging techniques must be widely available, reproducible at different sites, stable over time, easy to perform, low in cost, and provide maximum patient comfort and compliance. This is distinct from the common focus of university research on cutting-edge technology, which may have only single-site applicability. Proper consideration of the factors relevant to multicenter research facilitate study start-up, accelerate patient, recruitment, decrease patient dropout, and minimize sources of variability that undermine measurement precision and statistical power.

A. Selecting and Qualifying Imaging Sites

Selecting imaging sites for a clinical trial is a complex matter. Important considerations include: (1) the type and quality of imaging equipment available at the site and its compatibility with the equipment at the other sites, (2) the competence, motivation, reliability, and clinical-trials experience of the imaging technologists at the sites, (3) proximity to desirable clinical investigators, (4) patient convenience, (5) availability and ease of scheduling, as trial imaging competes with clinical-imaging slots, (6) process for transferring images between the site and the central radiology service, and (7) the cost of imaging. These factors must further be considered in light of the specific scientific, regulatory, and marketing requirements of the trial, the proposed method of image analysis, and the competence, experience, and systems compatibility of the central radiology service that will supervise the imaging, manage the image data, and perform the image analyses.

The degree to which imaging equipment across the multiple sites included in a clinical trial must be standardized depends on the type of measurements that will be made, and the ability of the central radiology service to deal with multivendor image data. Ideally, all equipment, software platforms, and upgrade schedules at all of the sites throughout the duration of the study should be identical, but this is rarely feasible. Knowing what deviations from this ideal can be tolerated without compromise to the scientific integrity of the study requires considerable sophistication and experience. In almost all cases, however, the imaging modality (plain radiography, CT, MRI, ultrasound, etc,) used to measure the particular morphological, compositional, or process-related feature of interest must be the same across all sites in the study. Additionally, the same specific instrument should be used for all serial examinations of an individual subject. Sometimes, however, equipment changes between visits are unavoidable. This can result in variations in geometrical magnification in radiography, bone mineral density measurements in DXA, signal and spatial homogeneity in MRI, and a variety of other parameters that can affect the study results. In such cases, specialized phantoms may be able to correct the technical variations. Phantoms can also be used to cross-calibrate different imaging devices at different sites. This enables the use of multivendor imaging equipment and thereby increase the pool of imaging sites applicable to a study. Numerous calibration phantoms for DXA are commercially acailable and routinely used in clinical trials of osteoporosis (Fig. 19), but special-purpose phantoms for other diseases are harder to find. Phantoms are also used to monitor the stability of imaging equipment over time (Fig. 20). Deviations in performance detected with these IAAs can be corrected by feedback to the imaging site or occasionally by postprocessing the image with specialized corrective software using quantitative information provided by the phantom.

How the data will be transferred to the radiology service is also an important consideration. The cost of mailing hard-copy images can be high and scales with the number of sites included. In contrast, unit cost of electronically transferring electronic images decreases with the number of sites networked. Only a few sites around world, however, are currently networked properly to allow electronic image transfers, but the number is increasing rapidly as teleradiology becomes more widespread. Alternatively, electronic image data can be transferred on a variety of inexpensive and relatively high storage capacity media, such as DAT tape or optical disc. Despite the fact that most imaging devices are currently formatted according to the DICOM 3.0 standard, managing and analyzing multivendor image data still requires specialized software. For example, files may be stored on digital linear tapes in DICOM 3.0 format, but the media itself may be proprietary.

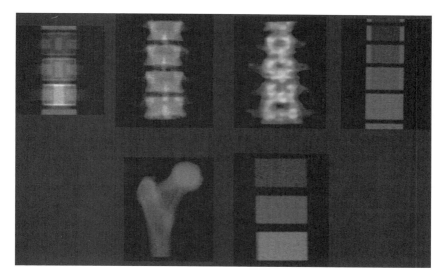

Figure 19 Various phantoms used for DXA scanner cross-calibration and quality control. Top row, left to right: European Spine Phantom (ESP), Hologic, Norland, Lunar Spine Phantoms. Bottom row: Hologic Hip and Hologic Block Phantoms.

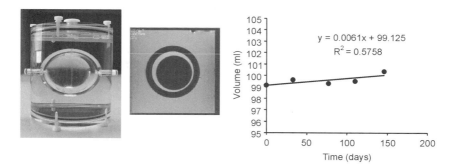

Figure 20 Phantom system for cartilage MRI. MR acquisition of articular cartilage in the knee and hip in longitudinal multicenter clinical trials is aided by the use of this phantom system (Crescent™, Synarc). These phantoms provide a variety of internal and external standard references for documenting and correcting longitudinal fluctuations in MRI hardware performance that may affect cartilage volume, thickness, and T2 measurements over time. With use of these phantoms, spatial drift of $<1\%$ over time is attainable. (Courtesy of Synarc, Inc., with permission.)

Motivated, competent imaging technologists, with direct experience conducting clinical trials are essential to the success of any study. Since imaging for clinical trials will always be only a marginal interest to mainstream radiology practice, clinical-imaging sites usually will not support protocols that deviate substantially from their clinical routines. Accordingly, the bigger the study, the simpler and more familiar the imaging protocol must be. Additionally, in almost all cases the imaging technologists must be given special training in how to perform the imaging protocol properly. This can be accomplished through either individual-site training or centralized training sessions at investigator meetings, and supplemented with detailed study manuals, videos, and/or interactive web-based instructional programs. Test runs of the imaging protocol and use of IAAs that simplify the technologists' work can also be extremely helpful. These IAAs include not only phantoms for calibrating and potentially correcting image quality, but positioning devices that ensure reproducible imaging on serial examinations (Figs. 17–20).

Often the proximity of the imaging site to a desirable clinical investigator is a critical factor. Convenience for the patient is also important. Scheduling imaging time for clinical trials can be difficult, as research cases compete with clinical cases, and it is the latter that are the sites' main priority. Poor availability of imaging time slots can slow a study considerably, and repeated cancellation and rescheduling of study patients can slow a study considerably, and repeated cancellation and rescheduling of study patients can lead to patient dropout. In some studies, imaging hubs fed by several clinical recruitment sites are used to reduce the total number of imaging sites needed. This can reduce cost and variability, and occasionally elevate the degree of protocol complexity that can be supported, but must be balanced carefully against patient inconvenience and scheduling capacity problems.

The cost of imaging can vary considerably from site to site and from country to country. International currency exchange rates are an important consideration in global trials. Additionally, some sites offer unit pricing, while others charge according to imaging time used. Repeat examinations necessitated by protocol violations or poor image quality, regardless of whether they were the result of inadequate training or lack of competence by the site, can add significant cost to a study. Accordingly, imaging cost must not be considered in isolation of the competence, experience, reliability and convenience of the imaging site in question.

B. Image Quality Control and Data Management

Once the imaging sites have been selected, the imaging protocol designed, and the technologists trained, image acquisition, transfer, and quality must be closely supervised to ensure a high-quality image set for analysis. Variability in image quality can be introduced either by the manner in which the subject is prepared for the examination and/or by improper calibration and maintenance of the imaging system. Device performance is documented and maintained by performing device quality control. One aspect of this is done by the imaging sites as part of their routine clinical quality control, but additional study-specific quality control must also be performed by the central radiology service. As mentioned above, this often requires the use of specialized phantoms (Figs. 19 and 20).

After the images are acquired according to the study-specific protocol, they are transmitted to the central radiologist(s) for review of protocol compliance, patient positioning, anatomical coverage, and image quality. This requires explicit image-quality (IQ) criteria. For example, serial radiographs of the knee for measuring joint-space narrowing (a marker of articular cartilage loss) in osteoarthritis must show reproducible projection of the anatomy, especially the region of the joint space to be measured. This is done both subjectively by an experienced reader visually comparing serial radiographs (Fig. 21) and by quantitative measurements of anatomical landmarks. For example, the distance between the anterior and posterior rims of the tibial plateau or between the superior margin of the patella and

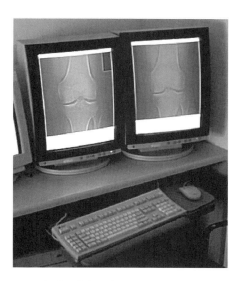

Figure 21 Quality control of serial radiography of the knee in osteoarthritis. Quality control includes assessment of the reproducibility of radioanatomical positioning. This is best done through side-by-side comparison of serially acquired images. (Courtesy of Synarc, Inc., with permission.)

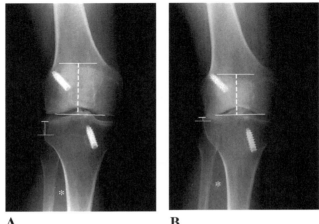

A   **B**

Figure 22 Evaluating serial knee positioning. Lateral joint-space width is markedly more narrowed in (B) than it is in (A) in this patient with prior anterior cruciate ligament repair. However, this difference is due to greater flexion of the knee in (B) rather than actual cartilage thinning, as the images were acquired only minutes apart. Note the lower position of the patella relative to the articular surface of the femur on (B) compared to (A). This, along with narrowing of the distance between the apex of the fibula and the tibial plateau are indicative of increased flexion in B. Image B also shows greater external rotation of the knee, as indicated by the widening of the interosseous space between the fibula and tibia (∗).

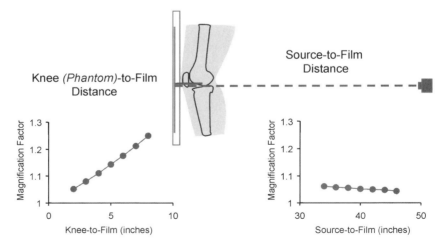

Figure 23 Magnification error in knee radiography. The graphs show how variations in the distance from the x-ray source to the film (right graph) have only minor effect on magnification, but even small changes in knee-to-film distance (left graph) alter magnification markedly.

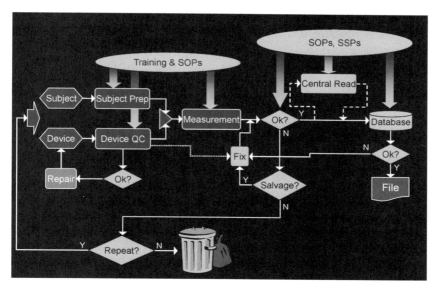

Figure 24 Image quality control in clinical trials. Shown is the process that a centralized clinical-trials radiology service uses to quality-control and analyze multisite image data.

the articular surface of the femur are measures of knee flexion, whereas the width of the interosseous space between the fibula and tibia is a marker of knee rotation (Fig. 22). Fiduciary phantoms (Fig. 17) and integrated image analysis software can also be used to verify knee side (right, left) and x-ray beam angulation on serial examinations to and quantify any changes in geometrical magnification (Fig. 23).

If the images are of acceptable quality, they are entered into the central study database. All processes performed by the central radiology service must be done in strict accordance with standard operating procedures (SOPs) and study-specific procedures (SSPs). If the image data do not pass the incoming quality inspection, a decision needs to be a made as to whether or not the data can be corrected, for example, by using information obtained from the instrument quality control. If the images cannot be salvaged, the imaging must be repeated or the data point discarded. The criteria applied are again controlled by SOPs, and data exclusions must be carefully documented. If repeat imaging is necessary a feedback loop must be designed with appropriate timing criteria. Final data consistency checks are applied before the data are submitted for filing and there may be an opportunity for a final adjustment if supported by appropriate instrument quality control information. The ultimate result of this process (Fig. 24) is a high-quality image set that conforms to rigorous quality assurance principles and can support high-quality image analysis.

VI. CENTRALIZED IMAGE ANALYSIS

In contrast to image acquisition, which must be performed at multiple sites, image analysis for clinical trails can be centralized (see above). This allows the use of specially trained readers and dedicated software to achieve superior results to those that would be attainable with noncentralized analysis. Based on current regulatory recommendations (Section VIII B1, Draft Guidance, Medical Imaging Drugs) readers must be independent (not participating in the study, not affiliated with the sponsor, and not affiliated with the institutions at which the study was conducted) and blinded (unaware of treatment identity, unaware—or have limited awareness—of patient-specific clinical information or of the study protocol, and not familiar with the inclusion and exclusion criteria for patient selection specified in the protocol). Two or more readers are recommended, but this raises the cost of image analysis, and the minimum number necessary to cover the work effectively should be used whenever possible. If necessary, large reading loads can be divided among multiple readers to accelerate the rate of analysis, but it is important that the same reader analyze all images in an individual patient series. Additionally, each reader should analyze the images independently of the other blinded readers and of any on-site readings performed by the clinical investigators or radiologists at the imaging sites.

Consistency among the readers should be quantified using a statistic appropriate for the nature of the data. Typically, percentage agreement or kappa statistic is used for nominal/existential markers; percentage agreement, weighted kappa statistic, or nonparametric correlation indices (Spearman's rho, Kendal's tau) are used for ordinal, or ranked categorical markers; and Pearson correlation coefficient, coefficient of variation, or intraclass correlation coefficient is used for dimensional or continuous markers. Conformity of the readers' assessments with a gold-standard image set is also desirable. Proper reader validation includes both up-front and ongoing validation. Up-front validation includes verification of prerequisite background training and experience in image interpretation and any licensing considerations that might be necessary. Qualified readers are trained in the scoring procedure or quantification method using training images, and then tested with a validation set of gold-standard images for which the results are known. Ongoing validation involves integrating cases from the validation image set (reader conformity) and cases already read by the reader

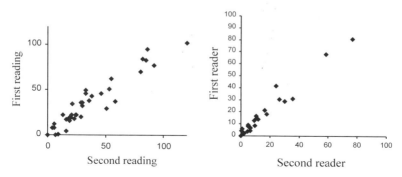

Figure 25 Reader reproducibility. (A) Duplicate reading by a single radiologist of 36 MRIs of osteoarthritic knees using whole-organ MRI scoring (WORMS). Pearson correlation coefficient = 0.95; coefficient of variation = 0.95; intraclass correlation coefficient = 0.9. (B) Results from different readers for modified-Sharp scoring of hand radiographs of patients with rheumatoid arthritis. Pearson's correlation coefficient = 0.96; Spearman's coefficient = 0.98.

(intrareader consistency) and by the other readers (interreader consistency) into the study reading set (Fig. 25).

The results of these validation exercises reflect the true study-specific performance of the readings and are a key metric of the central radiology service. Reading performance, however, depends on more than simply the reader's talent, training, and experience in clinical-trials radiology, although these factors are critical. The image-processing and analysis software used by the readers, particularly its user interface, also has a dramatic impact. In computer-assisted image reading, the reader is simply one component of an *integrated* reading *process*, and the consistency and conformity metrics described above therefore apply to the reading system as a whole, not the reader in isolation. Designing image analysis interfaces from the perspective of the end user, i.e., the reader, maximizes reading performance in terms of accuracy, consistency, speed, and capacity.

Clinical-trials reading systems include high-resolution monitors configured to allow side-by-side display of serially acquired images (Fig. 26). Images are typically stripped of any treatment or clinical information and presented in random chronological order to maintain reader *blindness* (admittedly, a poor choice of terms). Being able to compare serial images directly with each other increases readers' sensitivity for detecting interval change and their ability to adjust for subtle variations in anatomical positioning and projection. In studies in which the cases are blinded to chronological order, interim readings necessitate repeating these readings during the final assessment. So, side-by-side analysis of

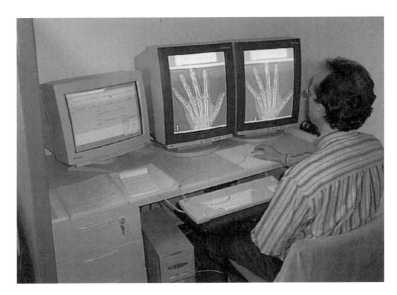

Figure 26 Image analysis workstation displaying digital images from serial examinations. Images are blinded to treatment and chronological order, edge-enhanced, and read in side-by-side comparison. This is combined with special measurement and reporting tools and automatic databasing of the results.

the images of a three-time-point trial that is read at the end of the study involves three readings. If an interim analysis is done at the second time point, the total number of readings will be five (visits 1 and 2 during the interim reading, and visits 1, 2, and 3 during the final reading).

In contrast to projectional images, such as radiographs and conventional scintigraphy, which present only one to four images per time point, tomographic imaging techniques, such as CT and MRI, typically yield dozens of individual cross-sectional images per scan. A two-time-point MRI series may comprise more than 150 images per patient. To cope with this volume of image data, all the sections of an individual scan are stacked in a single window of the workstation. Serial examinations are viewed in adjacent windows following anatomical registration to allow direct, side-by-side comparisons while synchronously scrolling rapidly back and forth through the anatomy (Fig. 27). This approach allows the reader to synthesize the cross-sectional information into a three-dimensional anatomical frame of reference, and is ideal for complex gestalt scoring methods, such as erosion scoring of rheumatoid arthritis with CT or MRI. In addition to multiple sections and multiple time points, readers of MRI must cope with multiple pulse sequences, each of which offers unique tissue

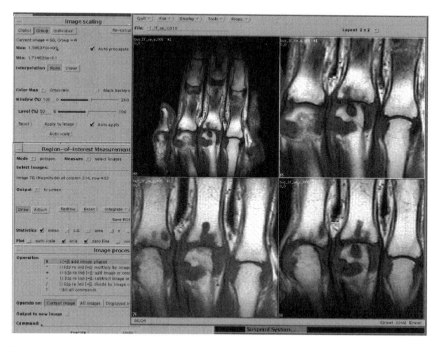

Figure 27 Specialized workstation for serial tomographic image analysis. Four windows are shown. The top right window and the two bottom windows each contain stacked coronal MRI images of the same wrist from one of three time points in a study. Images from all three time points are viewed together to improve the assessment of interval change, but the reader is blinded to the order in which the images were acquired. The images in these windows are magnified to facilitate reading. The top left window contains a nonmagnified image from one of the three datasets for anatomical reference. This figure does not illustrate the added dimension offered by scrolling through the sectional anatomy, as that would require animation. (From Ref. 12.)

information. With use of sophisticated multispectral and multitemporal data fusion techniques, all of this information can be combined into single cross-sectional images that are color-coded for different tissues and for the direction of change in these tissues over time (Fig. 28). In a similar approach, unique compositional or process-related information form one technology, such as PET or scintigraphy, can be combined with high-resolution anatomical information from another, such as CT or MRI, to create multimodality fusion images. A variety of complex image calculations can also be depicted in image mode (Figs. 7 and 29). These data-fusion and image-processing techniques expand the scope and speed of visual scoring in clinical trials.

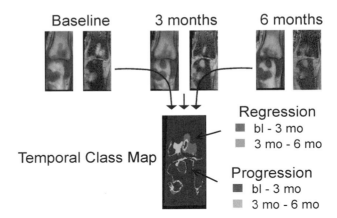

Figure 28   Multispectral and temporal data fusion. Fusion images generated from spatially registered MRI acquired with different pulse sequences and at different times display tissue classes and changes in different colors. This may facilitate image interpretation in longitudinal studies and provide an audit trail for regulatory purposes. (From Ref.12.) (See color insert.)

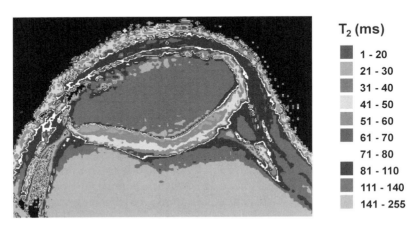

Figure 29   T2 relaxation map of normal adult articular cartilage. T2 map generated from multislice, multiecho (11 echoes: TE = 9,18,... 99 ms) spin echo images acquired at 3 T shows increasing T2 toward the articular surface. (Courtesy of B. J. Dardzinski, Ph.D., University of Cincinnati College of Medicine.) (See color insert.)

Computer-assisted image analysis also utilizes sophisticated quantification tools, to make planar, volumetric, and parametric measurements. These algorithms employ various degrees of automation, but virtually all require at least some reader interaction. In some cases this interaction can be minimal. For example, in computerized measurement of radiographic joint-space width in osteoarthritis, the algorithm automatically delineates the articular surface of the joint on digitized radiographs and measures the narrowest points (Fig. 30). However, an experienced radiologist must still verify that the computer traced the correct cortical margins (superimposition of overlapping tibial cortices can confuse even the best computers), and that the measurements were made within the weight-bearing regions of the joint and not, for example, between two marginal osteophytes. If the computer has made an error in judgment, the reader must correct the mistake using tools available on the workstation.

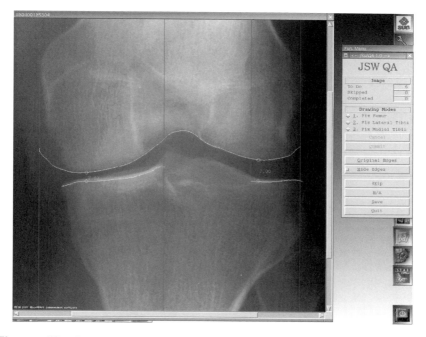

Figure 30 Computer-aided joint-space measurement. Specialized software automatically traces the articular surfaces of the femorotibial joints and calculates the minimum joint-space width or joint-space area. An experienced reader verifies the results and can adjust the horizontal limits of the sampled joint space to ensure that only the weight bearing regions are included. Additionally, the reader can correct any tracing errors, such as those associated with projectional superimposition of tibial cortices. (Courtesy of Synarc, Inc., with permission.)

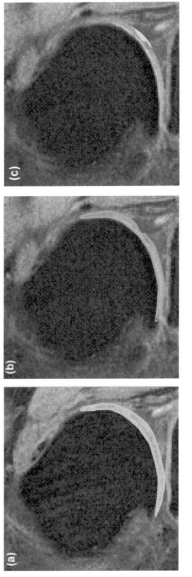

Figure 31 Regional cartilage volume quantification. Although the focal defect present at time point B represents less than 1% of the total femoral cartilage volume, by subtraction of the volume of only the slices containing the defect in B from the same slices on registered baseline images (A), the defect is quantified as 16% of regional femoral cartilage volume (C) (coefficient of variation = 1.2%). (From Ref. 22.)

In other cases, the reader must guide the computer more directly. For example, in computer-assisted cartilage volume measurement in osteoarthritis, small focal defects in the articular cartilage of the knee may be salient to the expert eye but measure less than 1% of the total femoral cartilage volume, and therefore be difficult to quantify. An alternative strategy is for the expert reader to identify all sections of the scan that contain a focal defect and instruct the computer to segment the cartilage in only those sections and the corresponding sections of spatially registered images from serial examinations [22]. The computer can then subtract the regional volumes and calculate the focal cartilage volumes loss directly (Fig. 31).

In addition to image processing, display, verification, and analysis, clinical-trials radiology workstations must also support electronic data entry, electronic signature, and automatic databasing of results. How the reader interfaces with these functions can dramatically affect the speed, accuracy, and capacity of the readings. The design of the score sheet is particularly important in this regard and requires an intimate understanding of how a reader approaches image analysis and data entry. A well-designed score sheet can reduce reader fatigue and human error in primary data entry. Designing the reader interface from the perspective of the reader is, therefore, critical to effective clinical-trials radiology. Ideally, reading system designers should be readers themselves. Another advantage of image analysis workstations is that they can provide an audit trail of the image measurements for scientific and regulatory purposes.

Accordingly, image analysis in clinical-trials is increasingly becoming a process in which expert readers and specialized software are tightly integrated and highly interdependent. The pace of development of usable, properly validated algorithms for this application, however, remains slow. This is principally because of the highly specialized nature and relatively small size of this niche market and its positioning upstream from clinical practice. Until a demand for these algorithms emerges in clinical practice, which for novel interests, such as cartilage quantification, will probably only follow the introduction of approved therapeutic agents linked to those markers, the bulk of these technical innovations are likely to remain prototypical, and few if any—outside those developed by commercial clinical-trials radiology services themselves—will be sufficiently validated to meet the regulatory requirements (21 CFR Part 11) in terms of electronic data integrity, safety, and traceability for use in clinical trials.

VII. CONCLUSION

Clinical-trials radiology is emerging as a unique and highly specialized application of medical imaging. Since it typically deals with innovative methods and regulatory and logistical demands that differ from those found in conventional clinical radiology, clinical-trials radiology requires special expertise, experience, and dedicated systems that cannot be found in mainstream clinical practice or conventional contract research services. However, when performed properly clinical-trials radiology can reduce uncertainty, time, and cost in global drug development, and thereby help bring new therapies into clinical use faster.

REFERENCES

1. Altman, R.D.; Hochberg, M.; Murphy, W.A.J.; Wolfe, F.; Lequesne, M. Atlas of individual radiographic features in osteoarthritis. Osteoarthritis Cart. **1996**, *3* (Suppl. A), 3–70.
2. Peterfy, C.G.; White, D.; Tirman, P.; et al. Whole-organ evaluation of the knee in osteoarthritis using MRI. European League Against Rheumatism, Glasgow, Scotland, 1999.
3. Wildy, K.; Zaim, S.; Peterfy, C.; Newman, B.; Kritchevasky, S.; Nevitt, M. Reliability of the whole-organ review MRI scoring (WORMS) method for knee osteoarthritis (OA) in a multicenter study. 65th Annual Scientific Meeting of the American College of Rheumatology, San Francisco, Nov 11–15, 2001.
4. Sharp, J.T.; Bluhm, G.B.; Brook, A.; et al. Reproducibility of multiple-observer scoring of radiologic abnormalities in the hands and wrists of patients with rheumatoid arthritis. Arthritis Rheum. **1985**, *28*, 16–23.
5. Genant, H.K. Methods of assessing radiographic change in rheumatoid arthritis. Am. J. Med. **1983**, *30* (Dec), 35–47.
6. Schuff, N.; Marmar, C.R.; Weiss, D.S.; et al. Reduced hippocampal volume and *n*-acetylaspartate in post traumatic stress disorder. Ann. N. Y. Acad. Sci. **1997**, *821*, 516–520.
7. Majumdar, S.; Weinstein, R.S.; Prasad, R.R. An evaluation of the fractal nature of trabecular bone. Medcine **1993**, *20*, 1611.
8. Weinstein, R.S.; Majumdar, S. Fractal geometry and vertebral compression fractures. J. Bone Miner. Res. **1994**, *9*, 1797–1802.
9. Xia, Y.; Farquhar, T.; Burton-Wurster, N.; Lust, G. Origin of cartilage laminae in MRI. JMRI **1997**, *7*, 887–894.
10. Dardizinski, B.; Mosher, T.; Li, S.; Van Slyke, M.; Smith, M. Spatial variation of T2 in human articular cartilage. Radiology **1997**, *205*, 546–550.
11. Erbel. The dawn of a new era in non-invasive coronary imaging. Herz **1996**, *21*, 75–77.
12. Peterfy, C.G. Magnetic resonance imaging of rheumatoid arthritis: the evolution of clinical applications through clinical trials. Semin. Arthritis Rheum. **2001**, *30*, 375–396.

13. Reddy, R.; Insko, E.K.; Noyszewski, E.A.; Dandora, R.; Kneeland, J.B.; Leigh, J.S. Sodium MRI of human articular cartilage in vivo. Magn. Reson. Med. **1998**, *39*, 697–701.

14. Bashir, A.; Gray, M.L.; Hartke, J.; Burstein, D. Nondestructive imaging of human cartilage glycosaminoglycan concentration by MRI. Magn. Reson. Med. **1999**, *41*, 857–865.

15. Bashir, A.; Gray, M.L.; Burstein, D. Gd-DTPA as a measure of cartilage degradation. Magn. Reson. Med. **1996**, *36*, 665–673.

16. Kusaka, Y.; Grunder, W.; Rumpel, H.; Dannhauer, K.-H.; Gersone, K. MR microimaging of articular cartilage and contrast enhancement by manganese ions. Magn. Reson. Med. **1992**, *24*, 137–148.

17. Shames, D.M.; Kuwatsuru, R.; Vexler, V.; Muhler, A.; Brasch, R.C. Measurment of capillary permeability to macromolecules by dynamic magnetic resonace imaging: a quantitative noninvasive technique. Magn. Reson. Med. **1993**, *29*, 616–622.

18. Brasch, R.C.; Weinmann, H.J.; Wesbey, G.E. Contrast-enhanced NMR imaging: animal studies using gadolinium-DTPA complex. Ann. J. Radiol. **1984**, *142*, 625–630.

19. van Dijke, C.; Kirk, B.; Peterfy, C.; Genant, H.; Brasch, R.; Kapila, S. Arthritic temporomandibular joint: correlation of macromolecular contrast-enhanced MR imaging parameters and histopathologic findings. Radiology **1997**, *204*, 825–832.

20. Cummings, S.R.; Black, D. Should perimenopausal women be screened for osteoporosis? Ann. Intern. Med. **1986**, *104*, 817–823.

21. Glüer, C.-C. Monitoring skeletal changes by radiological techniques. JBMR **1999**, *14*, 1952–1962.

22. Lynch, J.H.; Zarm, S.; Zhao, J.J.; Peterfy, C.G.; Genant, H.K.; Cartilage-lesion development following meniscal Surgery measured from knee MRI using image registration and segmentation. American College of Rheumatology, San Francisco, CA, 2001

4

Markers for Cardiac Repolarization and Risk Assessment

William J. Groh
Indiana University School of Medicine, Indianapolis, Indiana, U.S.A.

Gregory D. Sides
Eli Lilly and Company, Indianapolis, Indiana, U.S.A.

I. INTRODUCTION

The determination of whether an investigative pharmaceutical agent prolongs repolarization in the heart (an effect equivalent to prolongation of the QT interval on the surface electrocardiogram) has become a mandatory step in the drug development process. Such a determination takes considerable preclinical and clinical screening of cardiac biomarkers. This determination is rapidly evolving as an increasing body of knowledge is gained on the mechanisms by which drugs prolong cardiac repolarization and on identification of high-risk populations. Failure to recognize the potential of an investigative agent to prolong cardiac repolarization places future drug recipients at risk for life-threatening ventricular arrhythmias, particularly a distinctive polymorphic ventricular tachycardia termed "torsade de pointes." Appropriate market withdrawal of the drug may result. In this chapter, an approach to understanding the regulatory background, mechanisms, and the preclinical and clinical assessment of cardiac biomarkers in the determination of drug effects on cardiac repolarization is undertaken.

II. REGULATORY BACKGROUND

Recognition of the potential for drugs to prolong cardiac repolarization led the Committee for Proprietary Medicinal Products (CPMP) of the European Agency for the Evaluation of Medicinal Products (EMEA) to convene an ad hoc working group of experts in May 1996. This effort led to the publication in December 1997 of a seminal CPMP Points to Consider Document: "The Assessment of the Potential for QT Interval Prolongation by Non-Cardiovascular Medicinal Products" [1].

The CPMP document was followed in September 1999 by a draft position paper by Fenichel and Koerner of the U.S. Food and Drug Administration titled "Development of Drugs That Alter Ventricular Repolarization" [2].

In March 2001, the Therapeutic Products Directorate of Health Canada published a draft guidance document titled "Assessment of the QT Prolongation Potential of Non-Antiarrhythmic Drugs"[3].

The occurrence of torsade de pointes and sudden death in clinical studies of nonantiarrhythmic drugs is extremely rare. In addition, the background rate of torsade de pointes in the general population is difficult to establish. Whether QT interval prolongation predicts the potential for torsade de pointes or sudden death for nonantiarrhythmic drugs is not known. The use of QT interval prolongation as a predictor of torsade de pointes and sudden death is based on the class III antiarrhythmic registration databases (particularly sotalol). Thus, QT interval prolongation is considered a biomarker for torsade de pointes and sudden death. All drugs should be evaluated for possible effects on cardiac repolarization. Drugs that are naturally occurring proteins given at physiological doses as replacement therapy and blood products may be exempt from the requirement for cardiac repolarization assessment.

The benefit/risk profile of the drug will be considered at the time of regulatory agency review. Drugs that prolong the QT interval may be acceptable for approval if the drug provides a clinical benefit not provided by current therapy, provides a clinical benefit in treatment failures, or provides a long-term mortality benefit compared to current treatment.

III. MECHANISMS OF QT PROLONGATION AND TORSADE DE POINTES

A. The Surface ECG and the Cardiac Action Potential

The electrical activity of the heart controls the timing and propagation of cardiac mechanical activity (excitation–contraction coupling) [4]. The electrical state of cardiac myocytes varies from a resting negative potential (polarization) to a positive potential (depolarization) with these changes reflected in the action potential (Fig. 1). The surface electrocardiogram (ECG) is a voltage summation of the electrical activity of the underlying cardiac myocytes. The P wave reflects atrial depolarization, the QRS complex reflects ventricular depolarization, and the T wave ventricular repolarization (return to a polarized state). The QT interval, measured from the onset of the QRS complex to the termination of the T wave, is a voltage summation of the duration of the action potentials of ventricular myocytes.

The action potential is generated as a result of current flow across the cell membrane (sarcolemma). The current flow occurs secondary to changes in the membrane permeability of cardiac sarcolemmal ion channels. The direction and magnitude of current flow are determined by multiple complex factors including the transmembrane concentration gradient of the permeable ion and the voltage gradient, as reflected by the Nernst equation, and the unitary conductance and density of the ion channel [5]. Flow of positive ions into the myocyte or negative ions out of the myocyte, defined as an inward current, results in cell depolarization. The opposite, flow of positive ions out of the myocyte or negative ions into the myocyte, defined as an outward current, results in cell repolarization. Cardiac myocytes are electrically coupled to each other by gap junctions forming a synctium. This coupling minimizes the voltage differential between cells and allows the normal propagation of action potentials.

B. Prolonged Cardiac Repolarization

Ventricular repolarization occurs when increasing outward currents (primarily K^+) and decreasing inward currents (primarily Na^+ and Ca^{2+}) drive the membrane potential back to a resting negative state. A decrease in outward current or an increase in inward current during the plateau or repolarization phase of the action potential will result in prolongation of cardiac repolarization. Pharmacological agents that prolong repolarization do this by either blocking outward currents (most commonly) or potentiating inward currents (more rarely). The prolongation of the action potential duration results in QT-interval prolongation on the ECG. As will be discussed, the prolongation of the action

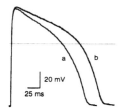

Figure 1 Action potential recorded from guinea pig papillary muscle using a microelectrode technique. The line is at a potential of 0 mV. The recordings were done in a control state (a) and after the addition of the class III antiarrhythmic drug, *d,l*-sotalol (b). Significant prolongation in the action potential is observed as an effect of the class III antiarrhythmic drug. (From Ref. 81.)

potential and QT interval can result in a pathophysiological state conducive to the generation of specific and serious ventricular arrhythmias.

An increasing understanding of the molecular basis and electrophysiology of the ion channels responsible for the action potential has occurred over the last decade. The number and molecular complexity of these ion channels makes a full review of this subject beyond the scope of this chapter. One channel, termed HERG, plays a vital role in normal human cardiac repolarization and is also a common target for drug blockade. An understanding of the HERG channel is essential to evaluating biomarkers in assessing cardiac repolarization.

C. HERG (Human Ether-à-Go-Go-Related Gene)

The delayed rectifier K^+ current, termed I_K, is a repolarizing outward current consisting of two distinct components, I_{Kr}, a rapid component, and I_{Ks}, a slow component [6]. The role of I_{Ks} in cardiac repolarization in humans remains controversial [7–10]. I_{Kr} plays an important role in action potential repolarization and in the normal rate-dependent shortening of the action potential. Virtually all pharmacological agents that prolong cardiac repolarization clinically do this, at least in part, by blockade of I_{Kr} [11]. In 1995, the human gene encoding I_{Kr} was identified [12]. This gene was a homolog of one initially cloned from a *Drosophila* mutant in which a dance-like movement disorder was observed on exposure to ether, hence the name human ether-à-go-go-related gene, or HERG [13]. Mutations in HERG were found to be linked and to encode at least some cases of the human disease, the congenital long QT syndrome (see below) [14]. The HERG protein can be expressed in heterologous cell systems and the effect of pharmacological agents on this channel can be determined [15]. The structure of the HERG channel may play a role in the diverse group of compounds found to block it [16]. Indeed, a description of the structure–activity relationship leading to drugs that prolonged repolarization predated the discovery of HERG [17]. In general, lipophilic compounds comprised of a tertiary amine linked via a highly variable chain to a para-substituted phenyl ring increase the likelihood of binding to and blocking of I_{Kr} (HERG). This pharmacophore is shared by many but not all compounds found to prolong repolarization [16].

D. Factors Modulating Ventricular Repolarization

Different regions of the ventricular myocardium have different properties of repolarization [18]. If a wedge of ventricular myocardium is cut in a transmural fashion, three regions, the endocardium, the midmyocardium, and the epicardium, can be defined. Electrophysiological evaluations of single cells from these different regions show significant variability in action potential duration. The midmyocardial or M cells have much longer action potential duration than either the endocardial or epicardial cells [19]. These M cells, along with Purkinje fiber cells, have the longest action potentials in the heart. These two cell types also prolong their action potential to a greater degree in response to drugs prolonging repolarization as compared to cells in other regions of the heart [11]. This can increase the heterogeneity in repolarization and, as will be subsequently discussed, increase the propensity for serious arrhythmias. The difference in action potential duration between cell types is diminished in vivo by the electrical coupling between cells [20].

Different clinical populations and individuals can have factors, both physiological and pathophysiological, that vary the susceptibility to prolonged ventricular repolarization. The congenital long-QT syndromes are inherited diseases in which abnormally prolonged repolarization is observed along with a propensity toward the characteristic arrhythmia of torsade de pointes [21–24]. Genetic linkage analysis and chromosome mapping have identified several ion channel targets responsible for these syndromes (Table 1). The ion channels involved can be the same channels that are targets for drug blockade. Although individuals with overt long-QT syndrome are rare, it is hypothesized that there are individuals with a subtler genetic predisposition to prolonged repolarization. These individuals have been described as having a forme fruste of a long-QT syndrome or having reduced cardiac repolarization reserve [11,25,26]. They may require only minor drug blockade of repolarizing outward currents to prolong repolarization from initially normal values and thus increase the risk of serious arrhythmias. This may be one reason why drugs that prolong the QT interval to only a minor degree in clinical trial evaluation have been associated with documented cases of torsade de pointes and sudden death.

The QT interval is longer in young and middle-aged women than men of a comparable age [27,28]. Female sex hormones have been found to prolong the QT interval and downregulate repolarizing K^+ channel expression [29]. Women with the congenital QT syndrome have longer QT intervals than men with the same syndrome and are more likely to suffer from associated arrhythmias [30,31]. Women are at an increased risk of excessive prolongation of the QT interval when given drugs prolonging repolarization [32]. In a rabbit model, treatment with the I_{Kr}-blocking agent erythromycin required almost a 10-fold increase in concentrations to prolong the QT interval to the same degree in males as females [33].

Cardiac hypertrophy and failure prolong the QT interval [34–37]. This results from a downregulation in several repolarizing currents [38]. Patients with severe congestive heart failure can develop excessive prolongation of repolarization and associated arrhythmias [39,40].

At slower heart rates the cardiac action potential duration normally increases. Excessive heart rate slowing such as seen in acquired heart block can

Table 1 Current Genetic Information in Long QT Syndrome

Chromosome locus	Gene	Current
Autosomal dominant (Romano-Ward)		
LQT1 11p15.5	KVLQT1 (KCNQ1)	↓ I_{Ks}
LQT2 7q35–36	HERG	↓ I_{Kr}
LQT3 3p21–24	SCN5A	↑ I_{Na}
LQT4 4q25–27	Unknown	Unknown
LQT5 21q22.1–22.2	MinK (KCNE1)	↓ I_{Ks}
LQT6 21q22.1–22.2	MiRP1 (KCNE2)	↓ I_{Kr}
LQT7 unknown	Unknown	Unknown
Autosomal recessive (Jervell-Lange-Nielsen)		
JLN1 11p15.5	KVLQT1 (KCNQ1)	↓↓ I_{Ks}
JLN2 21q22.1–22.2	MinK (KCNE1)	↓↓ I_{Ks}
JLN3 unknown	Unknown	Unknown

HERG = human "ether-a-go-go" related gene; I_{Kr} = rapidly activating component of delayed rectifier potassium current; I_{Ks} = slowly activating component of delayed rectifier potassium current; I_{Na} = sodium current; JLN = Jervell-Lange-Nielsen syndrome; MiRP1 = minK related peptide 1; SCN5A = cardiac voltage-dependent sodium channel gene; LQTS = Long QT Syndrome.

lead to a markedly prolonged QT interval and arrhythmias without any other trigger [41]. Drugs prolonging repolarization by blocking I_{Kr} have their greatest effect at slow heart rates [42]. This property, reverse use dependence, further increases the QT interval at slower heart rates and can increase the susceptibility to arrhythmias [43].

Hypokalemia reduces the magnitude of I_{Kr} and may enhance drug effects prolonging repolarization [44].

Interactive factors that diminish the metabolism of drugs prolonging repolarization will increase plasma concentrations and lead to a more prolonged QT interval. This can occur secondary to hepatic or renal dysfunction or related to concomitant medications interfering with metabolism. Two such targets appear to be the P450 isoenzymes CYP3A4 and CYP2D6 [18,45].

E. Drugs That Prolong Cardiac Repolarization

The property of a pharmaceutical agent to prolong the electrocardiographic QT interval (and the cardiac ventricular action potential) is termed a class III antiarrhythmic effect (after the Vaughn-Williams classification scheme) [46,47]. The list of drugs having the potential to prolong the QT interval is extensive. Such is the concern with the effect of drugs to prolong the QT interval that an international registry has been established to document cases of any drugs associated with prolonged QT interval and arrhythmia (website www.qtdrugs.org). In certain antiarrhythmic drugs, the ability to prolong cardiac repolarization is a sought-after therapeutic effect [47]. By prolonging repolarization these drugs increase the refractory period of cardiac muscle thereby making this tissue unexcitable for propagation of reentry-based arrhythmias. The class III antiarrhythmic drugs have shown variable efficacy in controlling life-threatening arrhythmias often with proarrhythmic potential limiting their use [48–50]. The more effective agents tend to have multiple antiarrythmic actions beyond a pure effect of prolonging repolarization. An example of such an agent is amiodarone with Na^+, Ca^{2+}, and beta-adrenergic blocking actions accompanying the class III antiarrhythmic effects [51].

There is no evidence that the therapeutic efficacy of the nonantiarrhythmic drugs that prolong repolarization is tied to that property. Thus, the class III antiarrhythmic effects of these drugs are an undesirable action. An example of this is the nonsedating H1-antihistaminic agent terfenadine. This drug was removed from the U.S. market when evidence of QT prolongation and associated arrhythmias were recognized [52]. The first metabolite of terfenadine, fexofenadine, retains antihistaminic effects with much less evidence of any action on cardiac repolarization [53]. The classes of nonantiarrhythmic drugs possessing the ability to prolong repolarization cross multiple therapeutic areas and includes psychiatric, antimicrobial, antihistamine, cardiac anti-ischemic/ vasodilator, and other miscellaneous agents. This wide spectrum of drugs that prolong cardiac repolarization has resulted in the necessity to test essentially all new drug applications for this untoward effect.

F. Arrhythmias Associated with Prolonged Ventricular Repolarization

Prolongation of the cardiac ventricular action potential, in itself, does not cause any harmful effects. Indeed, antiarrhythmic drugs with a pure class III effect can have a positive inotropic action on cardiac muscle and thus theoretically improve the hemodynamic status of individuals with congestive heart failure [54,55].

Unfortunately, a specific type of polymorphic ventricular tachycardia, termed torsade de pointes, can occur in the setting of prolongation of the QT interval [56]. Indeed, it is a requirement that prolongation of the QT interval is recognized at some point in order to label a polymorphic ventricular tachycardia torsade de pointes [37]. This arrhythmia, named by the French cardiologist Dessertenne, for the continuous twisting of its axis (Fig. 2) can lead to symptoms of palpitations and syncope and, if continuous or degenerating into ventricular fibrillation, sudden arrhythmic death [41,57]. Much is understood on how this distinctive ventricular tachyarrhythmia is generated from the underlying substrate of prolongation of cardiac repolarization. The prolongation of cardiac repolarization increases the vulnerability of the cardiac myocyte to depolarizing shifts in the membrane potential during the terminal phase of the action potential. These depolarizing shifts are related to reactivation of Ca^{2+} currents and/or activation of the Na^+-Ca^{2+} exchange current [11,58]. Because of the perturbation in the subtle balance between inward and outward currents caused by delayed repolarization, the depolarizing shifts in membrane potential can reach threshold and result in the generation of a second action potential. This action potential is termed an early afterdepolarization and is felt to be the initiating beat that can subsequently generate torsade de pointes [59]. Not all ventricular extrasystoles related to early afterdepolarizations will lead to torsade de pointes. Indeed, the recognition of new-onset ventricular extrasystoles commonly in a bigeminal pattern can herald the onset of torsade de pointes [41]. The M cells and Purkinje cells are the ventricular myocytes with the longest action potential duration at baseline and with the greatest vulnerability to further prolongation with class III antiarrhythmic action [58]. These cells are the most likely site at which early afterdepolarizations initiate and reach threshold. The likelihood of the onset of torsade de pointes is enhanced by the marked variability in the action potential duration between the M cells and endocardial and epicardial cells that occurs with class III antiarrhythmic effects [19,60]. Torsade de pointes is a reentry-based arrhythmia and this heterogeneity (dispersion) in repolarization is the appropriate milieu for its generation and continuation [61,62]. The dispersion in refractoriness can be measured clinically by determining the greatest difference in the QT interval measured on each lead of an ECG. The normal QT dispersion of less than 50 ms is increased by drugs that prolong repolarization and often will be further increased prior to episodes of torsade de pointes [63]. Whether QT dispersion provides any additional information beyond the accurate measurement of the QT interval in determining risk for torsade de pointes is debated [11].

The incidence of recognized torsade de pointes in individuals receiving drugs with class III action is dependent on the type of agent. In general, drugs employed as antiarrhythmics have the highest overall incidence. Torsade de pointes occurs in 1–10% of individuals treated with quinidine, *d,l*-sotalol, dofetilide, or ibutilide [11,55,56,64–67]. With the exception of quinidine, the incidence of torsade de pointes with these agents is dose-dependent. The incidence of torsade de pointes with amiodarone, a commonly employed antiarrythmic drug, is much lower, estimated at less than 1% [51,66,68]. This low incidence is likely related to the Ca^{2+}-channel-blocking effect of amiodarone that can diminish the propensity toward early afterdepolarizations. The incidence of torsade de pointes with nonantiarrhythmic drugs found to prolong repolarization is low. However, this incidence is difficult to estimate and is underreported. The incidence of torsade de pointes with cisapride, a gastrointestinal prokinetic agent that blocks HERG, is estimated at 1 in 120,000 individuals treated [69]. In cases of torsade de pointes observed in individuals with treatment with

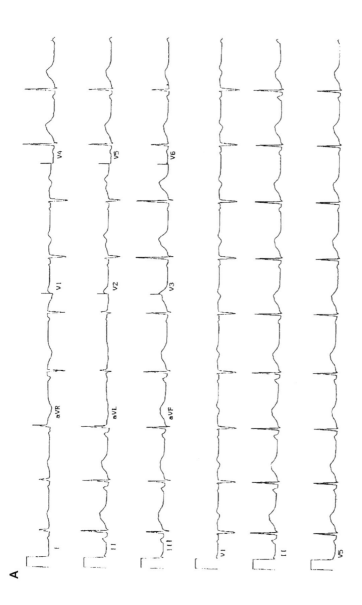

Figure 2 Electrocardiographic findings in torsade de pointes. Twelve-lead electrocardiogram taken on day 4 after initiation of the class III antiarrhythmic drug *d,l*-sotalol (A). Note the prolonged QT interval. Torsade de pointes with onset occurring during ventricular bigeminy (B). Note the rapid change in axis. Twelve-lead electrocardiogram taken on day 2 after cessation of *d,l*-sotalol (C). The QT interval has normalized.

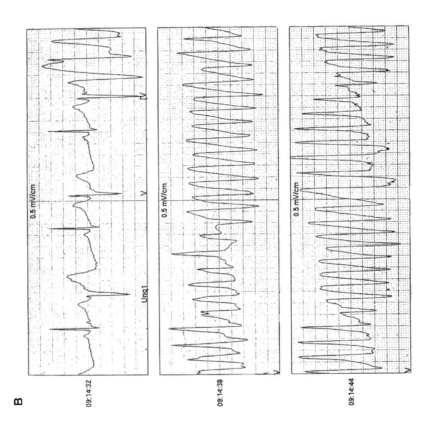

Figure 2 continued.

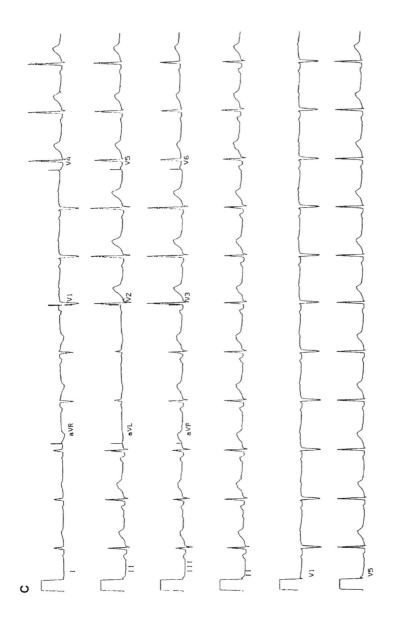

Figure 2 continued.

nonantiarrhythmic drugs, there are often found to be multiple offending agents, high dosages used, or situations limiting normal metabolism.

The factors as described above that increase the likelihood of prolonged repolarization in individuals receiving drugs with class III action also increase the likelihood of torsade de pointes. For example, there is a two-to-threefold higher incidence of torsade de pointes in women treated with antiarrhythmic drugs compared to men [32,65]. Bradycardia, and, more typically, heart rate pauses, commonly precede the onset of this ventricular tachycardia. This so-called pause dependence is a hallmark of drug-induced torsade de pointes [41]. This is likely related to an exaggeration of the normal increase in action potential duration seen with slower heart rates. It may be a similar rate-slowing mechanism responsible for the increased incidence of torsade de pointes at the time of conversion from atrial fibrillation to sinus rhythm [70]. The degree of QT-interval prolongation on the ECG that precedes torsade de pointes is highly variable. Some authors have proposed uncorrected QT intervals of 550–600 ms as a value that might frequently herald impending torsade de pointes [37].

Unfortunately, many cases occur with QT intervals much closer to the normal range and many individuals show little or no QT-interval prolongation when an ECG is analyzed during normal sinus rhythm [41]. The difficulty in determining a QT-interval value likely to predict torsade de pointes may relate as well to the presence of confounding U waves making accurate and uniform measurement of the QT interval problematic. U waves are commonly present in the setting of prolonged repolarization. They likely originate in areas of ventricular muscle in which delay in repolarization is the greatest, the M cells. The presence of large U waves, especially pause-dependent U-wave accentuation and new postpause ventricular extrasystoles, has been reported to be the most sensitive herald of subsequent torsade de pointes [41].

Torsade de pointes tends to be a paroxysmal ventricular tachycardia with short bursts of rapid rhythm. This often allows the monitored patient to be treated prior to a fatal outcome. Treatment options work by shortening the QT interval and preventing heart rate pauses (isoproterenol infusion or temporary pacing) or diminishing early afterdepolarizations (intravenous magnesium or beta-adrenergic blockers) [37]. External defibrillation may be required in individuals in whom a sustained arrhythmia occurs. Any potential offending drugs should be withdrawn and serum potassium levels should be monitored and kept in the high-normal range.

IV. PRECLINICAL ASSESSMENT EVALUATING DRUG EFFECT ON REPOLARIZATION

Both in vitro and in vivo methods provide biomarkers assessing the potential of a drug to prolong repolarization in humans. These methods use standard basic electrophysiology techniques to assess drug effects on individual ion channels, action potential morphology and duration, the animal ECG, and the ability to induce the specific arrhythmias associated with prolonged repolarization. When these preclinical assessments are undertaken, it is important that both the parent compound and metabolites be evaluated. This can require reassessment based on knowledge of further metabolites. The concentrations tested should extend beyond the therapeutic range of the drug with a typical recommendation of a total of 3-log-units concentrations. Testing multiple concentrations over a wide range will also help to distinguish spurious from true results in that a concentration-dependent effect on the specific biomarker should be anticipated. If drug effect on a particular biomarker is identified, a 50% inhibitory concentration (IC_{50}) value should be determined from the concentration–effect curve. An understanding of the solubility of the drug under study is necessary to assure that high concentrations can be tested. Drugs can adhere to the glass or plastic of perfusion systems and therefore postperfusion concentrations should be assessed. Studies should be run at as physiological a temperature as possible. In all test methods, control compounds, known to prolong repolarization, should also be assessed. These control compounds are typically I_{Kr} blockers. The choice of animal species for cells or tissue should be carefully considered to assure that ion channels and action potentials are similar to those found in humans. Common models chosen include dog, swine, rabbit, and guinea pig. Action potential morphology and duration and repolarization are so different in mice and rats that these species are generally not used. Human cells and tissue, commonly of atrial origin, also are available for testing [71,72]. The gender of the animals studied can vary the results and should be recorded [73]. The results of all the testing methods should be evaluated as a whole to determine drug effects on repolarization. Inconsistent results between testing methods may require further careful evaluation.

A. In Vitro Ion Channel Assessment

The effect of a drug on individual ion channels (currents) can be studied using voltage clamp techniques. These studies are done in enzymatically dissociated myocardial cells studied acutely or after culture and in heterologous expression systems in which the ion channel protein of interest is expressed (Fig. 3). The heterologous expression systems employ cells that have little intrinsic voltage-activated channel activity including *Xenopus laevis* oocytes and mammalian cell lines such as human embryonic kidney cells (HEK293) and Chinese hamster ovary (CHO) cells. The expression of cloned ion channel protein and voltage clamp study of the large *Xenopus* oocytes is easier than that in mammalian cell lines. However, the overall large size and presence of the yolk material in oocytes make study of rapidly activated channels and lipophilic drugs difficult. Because of the importance of the HERG protein in drug-mediated effects on cardiac

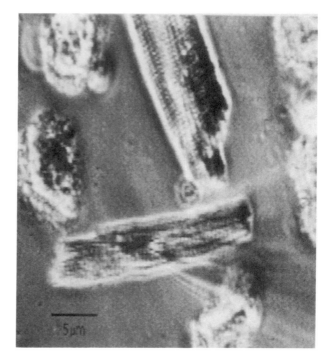

Figure 3   Photomicrograph of an isolated guinea pig cardiac ventricular myocyte during voltage clamp studies. Note the glass electrode impaling the bottom cell.

repolarization, study of this expressed channel is key. In many recent publications the effect of drugs known to prolong cardiac repolarization have been quantitated in HERG-transfected cells [72,74–77].

Study of native ion currents in myocardial cells allows the evaluation of a wide range of currents including Na^+, Ca^{2+}, and K^+ currents. These currents all have the potential to modulate repolarization and thus require evaluation [11].

B.   In Vitro Action Potential and Electrocardiographic Assessment

Dissociated myocardial cells studied using voltage-clamp techniques are not an adequate model for the evaluation of drug effects on action potential morphology or duration. The action potential is altered by the requisite perfusing and pipette solutions needed to keep cells viable and often varies significantly with

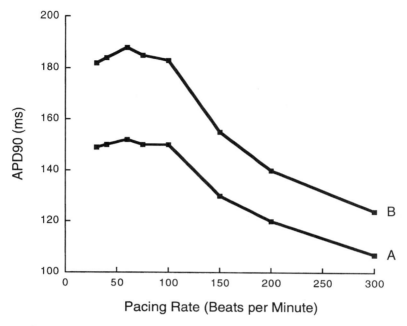

Figure 4 Action potential duration at 90% repolarization (APD90) as a function of pacing rate in guinea pig papillary muscles measured using a microelectrode technique. In a control state, the APD90 is rate-dependent, decreasing as pacing rate is increased (A). With the addition of 10 μM E-4031, a selective I_{Kr} blocker, the APD90 prolongs (B). APD90 prolongation is greater at slower rates (reverse use dependence). (Adapted from Ref. 42.)

the duration of the voltage clamp [5]. Isolated multicellular tissue samples, such as Purkinje fibers, papillary muscles, and myocardial wedge slices, can be studied using a microelectrode technique and provide a more reproducible and consistent action potential recording. These studies should be done with a sufficient period of time between changes in drug concentrations to assure a steady-state condition. The tissue should be paced at multiple rates to assess rate dependence of any drug effect (Fig. 4). Drugs with class III antiarrhythmic effects due to I_{Kr} blockade prolong action potential duration to a greater extent at slower rates (reverse use dependence). Observing such an effect can assist in the understanding of a drug effect on repolarization. The use of tissue with the greatest susceptibility to action potential prolongation can aid in observing a more limited, but important, repolarization effect. For this reason, study of midmyocardial wedge tissue may be useful [78]. Drug effect is most typically quantitated using action potential duration at 90% repolarization (APD$_{90}$).

Intact, isolated, Langendorff-perfused hearts with recordings of surface ECG or action potential can be used to screen drugs for repolarization effects. Typical animal species employed are guinea pig and rabbits.

If a drug is found to prolong action potential duration in isolated cardiac tissue or whole, perfused hearts in a concentration-dependent manner, it is reasonable to expect prolongation of repolarization in humans.

C. In Vivo Action Potential and Electrocardiographic Assessment

In vivo studies in both conscious and anesthetized animals provide evaluation for multiple cardiovascular and noncardiovascular untoward effects of an investigational drug. For evaluation of effects on cardiac repolarization an appropriate animal species should be selected, the anesthetic agent (if used) should be free of any effect on repolarization, and conscious animals should be unrestrained to assure a normal sympathetic tone. Radiotelemetry devices may be employed. Measurement of the QT interval should be done from a minimum of three successive beats and averaged. Correction of the QT interval for rate may require individualized assessment since formulations used in humans typically do not apply. Pacing at a constant rate is another option.

D. In Vitro and In Vivo Arrhythmia Models

Early afterdepolarizations and short runs of triggered beats can be observed in isolated tissues and Langendorff-perfused hearts on exposure to class III antiarrhythmic drugs [79]. The presence of these findings accompanying action potential or QT-interval prolongation in the evaluation of a drug would heighten the concern for a significant repolarization effect. Recent publications have detailed animal models in which pathophysiological triggers have been employed to increase the propensity toward arrhythmias associated with drugs that prolong repolarization. Models include alpha-choralose-anesthetized rabbits treated with the alpha-1 agonist methoxamine and a dog model using a trigger of bradycardia and hypokalemia or atrioventricular block and short/long/short pacing [61,80].

V. CLINICAL ASSESSMENT

Incorporating the preclinical and clinical literature and regulatory thought into the performance of electrocardiograms in clinical studies leads to the following considerations.

ECGs should be collected in phase I single- and multiple-dose safety study(ies) and in the pivotal phase II study.

ECGs should be obtained from a minimum of 100 subjects for drugs that have no appreciable in vitro HERG effect and do not prolong the QT interval in a large animal model (dog or monkey).

ECGs should be obtained from a minimum of 200 subjects for drugs that block HERG at clinically relevant drug concentrations or prolong the QT interval in the dog or monkey. Clinically relevant in vitro HERG blockade is thought to occur when ≥ 20% inhibition of the HERG channel is observed at concentrations equivalent to plasma or myocardial tissue free drug concentrations of the parent drug and/or principal metabolites. Therefore, protein binding of both parent and metabolite should be determined prior to first human dose. If HERG blockade or QT prolongation in animals occurs, then electrocardiograms will likely need to be collected in phase III trials as well as phase I and II.

The phase I single- and multiple-dose safety study(ies) should explore as large a dose range as possible based on the anticipated human dose, subject tolerability, and the no-observed-adverse-effect level (NOAEL) in animal toxicology studies. The dose range explored should be multiples above the anticipated therapeutic dose. ECGs in humans should be obtained concurrently with plasma drug concentrations at multiple time points. A standard procedure should be deployed in phase I studies for ECG collection, including lead placement, supine position, minimization of subject movement, and collection of resting 12-lead electrocardiograms after a few minutes of subject rest. Multiple baseline electrocardiograms should be obtained 1 or more days prior to study drug administration at the same time points planned for the dosing day.

ECGs for QT-interval assessment can be collected as either standard 12-lead ECGs, Holter ECG recording, or continuous ECGs collected with an electronic recorder (e.g., GE Marquette SEER) that can be programmed to collect 12-lead ECGs repetitively over a 24-h period.

A time course of ECGs should be performed. ECGs should be obtained at the following time points.

Immediately prior to study drug administration
Peak drug concentration of parent
Peak drug concentration of principal metabolites
Peak drug pharmacodynamic effect
Steady state
Trough drug concentration.

More frequent collection of ECGs should be considered for drugs that exhibit ≥ 20% inhibition of HERG or demonstrate QT-interval prolongation in animals (particularly a dose–response relationship between the QT interval and dose). In addition, a patient management algorithm and stopping rule should be included in the study design to ensure subject safety.

In the phase II trial, patients should have ECGs obtained at baseline, at time points surrounding the time of maximal drug concentration of parent and principal metabolites, and at trough drug concentration.

If the study drug is an inhibitor or substrate of a cytochrome P450 isoenzyme at clinically relevant drug concentrations and pharmacokinetic/pharmacodynamic drug interaction studies are planned, consideration should be given for collection of ECGs in these studies as well. ECGs should be obtained concurrently with plasma drug concentrations of the study drug and the other drug.

ECGs should also be obtained in pharmacokinetic studies in special populations to determine whether age, gender, renal, or hepatic impairment might be associated with a greater predisposition to delayed cardiac repolarization.

If HERG, animal, or human phase I and II data do not suggest the potential for the study drug to be associated with prolonged cardiac repolarization, then further collection of ECGs in phase III studies is likely not required for QT assessment.

A well-defined methodology should be employed for QT-interval measurement. Computerized QT-interval measurement should be supplemented with human overread of the QT interval. Human overread of the QT interval is particularly important for ECGs where the presence of a U wave or an abnormal T wave interferes with the computer's ability to accurately determine the offset of the T wave. Human overread is also important for ECGs with noise or motion artifact.

The physiological variability in QT-interval duration should be considered. The variability in a dataset increases the more frequently ECGs are performed. The mean QTcB-interval variability was 76 ± 19 ms in healthy subjects in one study [82] and 117 ± 28 ms in healthy subjects in another study [83].

The adverse-event database should be carefully analyzed for the frequency and nature of syncope, dizziness, tachycardia, torsade de pointes, and sudden death; however, the frequency of these events is rare even for drugs that prolong the QT interval.

VI. SUMMARY

Drugs that prolong cardiac repolarization have been associated with the development of a polymorphic ventricular tachycardia termed torsade de pointes, a potentially life-threatening arrhythmia that has been implicated in the occurrence of sudden death.

Many of the drugs that alter ventricular repolarization were developed as antiarrhythmics, but many other noncardiac drugs were developed without an expected effect on repolarization. Thus, prolonged cardiac ventricular repolarization is one of the two most common reasons for withdrawal of previously approved drugs from market (hepatotoxicity is the other reason for withdrawal). Therefore, most drugs must be assessed for the potential to prolong cardiac repolarization through an integrated in vitro, animal, and human approach.

REFERENCES

1. EMEA. The European Agency for the Evaluation of Medicinal Products. Human Medicines Evaluation Unit. Committee for Proprietary Medicinal Products (CPMP). Points to consider: the assessment of the potential for QT interval prolongation by non-cardiovascular medicinal products, CPMP/986/96. http://www.emea.eu.int/pdfs/human/swp/098696en.pdf (1997).
2. Fenichel, R.R.; Koerner, J. Development of drugs that alter ventricular repolarization (draft), 1999.
3. Therapeutic Products Directorate. Bureau of Pharmaceutical Assessment. Health Canada. Draft guidance document: assessment of the QT prolongation potential of non-antiarrhythmic drugs. http://www.hcsc.gc.ca/hpb-dgps/therapeut/htmleng/guidmain.html (2001).
4. Jalife, J.; Delmar, M.; Davidenko, J.M.; Anumonwo, J.M. *Basic Cardiac Electrophysiology for the Clinician*; Futura: Armonk, NY, 1999.
5. Hille, B. *Ionic Channels of Excitable Membranes*; Sinauer Associates: Sunderland, MA, 1992.
6. Sanguinetti, M.C.; Jurkiewicz, N.K. Two components of cardiac delayed rectifier K^+ current: differential sensitivity to block by class III antiarrhythmic agents. J. Gen. Physiol. **1990**, *96* (1), 195–215.
7. Li, G.R.; Feng, J.; Yue, L.; Carrier, M.; Nattel, S. Evidence for two components of delayed rectifier K^+ current in human ventricular myocytes. Circ. Res. **1996**, *78* (4), 689–696.
8. Iost, N.; Virag, L.; Opincariu, M.; Szecsi, J.; Varro, A.; Papp, J.G. Delayed rectifier potassium current in undiseased human ventricular myocytes. Cardiovasc. Res. **1998**, *40* (3), 508–515.
9. Veldkamp, M.W. Is the slowly activating component of the delayed rectifier current, IKs' absent from undiseased human ventricular myocardium? [editorial; comment]. Cardiovasc. Res. **1998**, *40* (3), 433–435.
10. Virag, L.; Iost, N.; Opincariu, M.; Szolnoky, J.; Szecsi, J.; Bogats, G.; Szenohradszky, P.; Varro, A.; Papp, J.G. The slow component of the delayed rectifier potassium current in undiseased human ventricular myocytes. Cardiovasc. Res. **2001**, *49* (4), 790–797.

11. Haverkamp, W.; Breithardt, G.; Camm, A.J.; Janse, M.J.; Rosen, M.R.; Antzelevitch, C.; Escande, D.; Franz, M.; Malik, M.; Moss, A.; Shah, R. The potential for QT prolongation and pro-arrhythmia by non-anti-arrhythmic drugs: clinical and regulatory implications. Report on a policy conference of the European society of cardiology. Cardiovasc. Res. **2000**, *47* (2), 219–233.

12. Trudeau, M.C.; Warmke, J.W.; Ganetzky, B.; Robertson, G.A. HERG, a human inward rectifier in the voltage-gated potassium channel family. Science **1995**, *269* (5220), 92–95.

13. Warmke, J.; Drysdale, R.; Ganetzky, B. A distinct potassium channel polypeptide encoded by the *Drosophila eag locus*. Science **1991**, *252* (5012), 1560–1562.

14. Curran, M.E.; Splawski, I.; Timothy, K.W.; Vincent, G.M.; Green, E.D.; Keating, M.T. A molecular basis for cardiac arrhythmia: HERG mutations cause long QT syndrome. Cell **1995**, *80* (5), 795–803.

15. Sanguinetti, M.C.; Jiang, C.; Curran, M.E.; Keating, M.T. A mechanistic link between an inherited and an acquired cardiac arrhythmia: HERG encodes the IKr potassium channel. Cell **1995**, *81* (2), 299–307.

16. De Ponti, F.; Poluzzi, E.; Montanaro, N. QT-interval prolongation by noncardiac drugs: lessons to be learned from recent experience. Eur. J. Clin. Pharmacol. **2000**, *56* (1), 1–18.

17. Morgan, T.K., Jr.; Sullivan, M.E. An overview of class III electrophysiological agents: a new generation of antiarrhythmic therapy. Prog. Med. Chem. **1992**, *29*, 65–108.

18. Camm, A.J.; Janse, M.J.; Roden, D.M.; Rosen, M.R.; Cinca, J.; Cobbe, S.M. Congenital and acquired long QT syndrome. Eur. Heart J. **2000**, *21* (15), 1232–1237.

19. Antzelevitch, C.; Shimizu, W.; Yan, G.X.; Sicouri, S.; Weissenburger, J.; Nesterenko, V.V.; Burashnikov, A.; Di Diego, J.; Saffitz, J.; Thomas, G.P. The M cell: its contribution to the ECG and to normal and abnormal electrical function of the heart [see comments]. J. Cardiovasc. Electrophysiol. **1999**, *10* (8), 1124–1152.

20. Anyukhovsky, E.P.; Sosunov, E.A.; Rosen, M.R. Regional differences in electrophysiological properties of epicardium, midmyocardium, and endocardium: in vitro and in vivo correlations. Circulation **1996**, *94* (8), 1981–1988.

21. Priori, S.G.; Barhanin, J.; Hauer, R.N.; Haverkamp, W.; Jongsma, H.J.; Kleber, A.G.; McKenna, W.J.; Roden, D.M.; Rudy, Y.; Schwartz, K.; Schwartz, P.J.; Towbin, J.A.; Wilde, A.M. Genetic and molecular basis of cardiac arrhythmias: impact on clinical management part III. Circulation **1999**, *99* (5), 674–681.

22. Priori, S.G.; Barhanin, J.; Hauer, R.N.; Haverkamp, W.; Jongsma, H.J.; Kleber, A.G.; McKenna, W.J.; Roden, D.M.; Rudy, Y.; Schwartz, K.; Schwartz, P.J.; Towbin, J.A.; Wilde, A.M. Genetic and molecular basis of cardiac arrhythmias: impact on clinical management parts I and II. Circulation **1999**, *99* (4), 518–528.

23. Roden, D.M.; Spooner, P.M. Inherited long QT syndromes: a paradigm for understanding arrhythmogenesis. J. Cardiovasc. Electrophysiol. **1999**, *10* (12), 1664–1683.

24. Chiang, C.E.; Roden, D.M. The long QT syndromes: genetic basis and clinical implications. J. Am. Coll. Cardiol. **2000**, *36* (1), 1–12.

25. Roden, D.M. Taking the "idio" out of "idiosyncratic": predicting torsades de pointes [editorial]. Pacing Clin. Electrophysiol. **1998**, *21* (5), 1029–1034.

26. Napolitano, C.; Schwartz, P.J.; Brown, A.M.; Ronchetti, E.; Bianchi, L.; Pinnavaia, A.; Acquaro, G.; Priori, S.G. Evidence for a cardiac ion channel mutation underlying drug-induced QT prolongation and life-threatening arrhythmias. J. Cardiovasc. Electrophysiol. **2000**, *11* (6), 691–696.

27. Rautaharju, P.M.; Zhou, S.H.; Wong, S.; Calhoun, H.P.; Berenson, G.S.; Prineas, R.; Davignon, A. Sex differences in the evolution of the electrocardiographic QT interval with age. Can. J. Cardiol. **1992**, *8* (7), 690–695.

28. Stramba-Badiale, M.; Locati, E.H.; Martinelli, A.; Courville, J.; Schwartz, P.J. Gender and the relationship between ventricular repolarization and cardiac cycle length during 24-h Holter recordings. Eur. Heart J. **1997**, *18* (6), 1000–1006.

29. Drici, M.D.; Burklow, T.R.; Haridasse, V.; Glazer, R.I.; Woosley, R.L. Sex hormones prolong the QT interval and downregulate potassium channel expression in the rabbit heart. Circulation **1996**, *94* (6), 1471–1474.

30. Lehmann,M.H.;Timothy,K.W.;Frankovich,D.;Fromm,B.S.;Keating,M.;Locati,E.H.; Taggart, R.T.; Towbin, J.A.; Moss, A.J.; Schwartz, P.J.; Vincent, G.M. Age–gender influence on the rate-corrected QT interval and the QT-heart rate relation in families with genotypically characterized long QT syndrome. J.Am.Coll.Cardiol.**1997**,*29*(1),93–99.

31. Locati, E.H.; Zareba, W.; Moss, A.J.; Schwartz, P.J.; Vincent, G.M.; Lehmann, M.H.; Towbin, J.A.; Priori, S.G.; Napolitano, C.; Robinson, J.L.; Andrews, M.; Timothy, K.; Hall, W.J. Age- and sex-related differences in clinical manifestations in patients with congenital long-QT syndrome: findings from the International LQTS Registry. Circulation **1998**, *97* (22), 2237–2244.

32. Makkar, R.R.; Fromm, B.S.; Steinman, R.T.; Meissner, M.D.; Lehmann, M.H. Female gender as a risk factor for torsades de pointes associated with cardiovascular drugs. J. Am. Med. Assoc. **1993**, *270* (21), 2590–2597.

33. Drici, M.D.; Knollmann, B.C.; Wang, W.X.; Woosley, R.L. Cardiac actions of erythromycin: influence of female sex. J. Am. Med. Assoc. **1998**, *280* (20), 1774–1776.

34. Beuckelmann, D.J.; Nabauer, M.; Erdmann, E. Intracellular calcium handling in isolated ventricular myocytes from patients with terminal heart failure. Circulation **1992**, *85* (3), 1046–1055.

35. Tomaselli, G.F.; Beuckelmann, D.J.; Calkins, H.G.; Berger, R.D.; Kessler, P.D.; Lawrence, J.H.; Kass, D.; Feldman, A.M.; Marban, E. Sudden cardiac death in heart failure: the role of abnormal repolarization. Circulation **1994**, *90* (5), 2534–2539.

36. Martin, A.B.; Garson, A., Jr.; Perry, J.C. Prolonged QT interval in hypertrophic and dilated cardiomyopathy in children. Am. Heart **1994**, *127* (1), 64–70.

37. Roden, D.M. A practical approach to torsade de pointes. Clin. Cardiol. **1997**, *20* (3), 285–290.

38. Tomaselli, G.F.; Marban, E. Electrophysiological remodeling in hypertrophy and heart failure. Cardiovasc. Res. **1999**, *42* (2), 270–283.

39. Moser, D.K.; Stevenson, W.G.; Woo, M.A.; Stevenson, L.W. Timing of sudden death in patients with heart failure. J. Am. Coll. Cardiol. **1994**, *24* (4), 963–967.

40. Stevenson, W.G.; Stevenson, L.W. Prevention of sudden death in heart failure. J. Cardiovasc. Electrophysiol. **2001**, *12* (1), 112–114.
41. Jackman, W.M.; Friday, K.J.; Anderson, J.L.; Aliot, E.M.; Clark, M.; Lazzara, R. The long QT syndromes: a critical review, new clinical observations and a unifying hypothesis. Prog. Cardiovasc. Dis. **1988**, *31* (2), 115–172.
42. Groh, W.J.; Gibson, K.J.; Maylie, J.G. Comparison of the rate-dependent properties of the class III antiarrhythmic agents azimilide (NE-10064) and E-4031: considerations on the mechanism of reverse rate-dependent action potential prolongation. J. Cardiovasc. Electrophysiol. **1997**, *8* (5), 529–536.
43. Hondeghem, L.M.; Snyders, D.J. Class III antiarrhythmic agents have a lot of potential but a long way to go: reduced effectiveness and dangers of reverse use dependence. Circulation **1990**, *81* (2), 686–690.
44. Sanguinetti, M.C.; Jurkiewicz, N.K. Role of external Ca^{2+} and K^+ in gating of cardiac delayed rectifier K^+ currents. Pflugers Arch. Eur. J. Physiol. **1992**, *420* (2), 180–186.
45. Priori, S.G. Exploring the hidden danger of noncardiac drugs [editorial; comment]. J. Cardiovasc. Electrophysiol. **1998**, *9* (10), 1114–1116.
46. Singh, B.N.; Vaughan Williams, E.M. A third class of anti-arrhythmic action. Effects on atrial and ventricular intracellular potentials, and other pharmacological actions on cardiac muscle, of MJ 1999 and AH 3474. Br. J. Pharmacol. **1970**, *39* (4), 675–687.
47. Zipes, D.P.; Jalife, J., Eds. *Cardiac Electrophysiology: from Cell to Bedside*; WB Saunders: Philadelphia, 2000.
48. Mason, J.W. A comparison of seven antiarrhythmic drugs in patients with ventricular tachyarrhythmias. Electrophysiologic study versus electrocardiographic monitoring investigators. N. Engl. J. Med. **1993**, *329* (7), 452–458.
49. Waldo, A.L.; Camm, A.J.; deRuyter, H.; Friedman, P.L.; MacNeil, D.J.; Pauls, J.F.; Pitt, B.; Pratt, C.M.; Schwartz, P.J.; Veltri, E.P. Effect of *d*-sotalol on mortality in patients with left ventricular dysfunction after recent and remote myocardial infarction. The SWORD investigators. Survival with oral *d*-sotalol. Lancet **1996**, *348* (9019), 7–12.
50. A comparison of antiarrhythmic-drug therapy with implantable defibrillators in patients resuscitated from near-fatal ventricular arrhythmias. The antiarrhythmics versus implantable defibrillators (AVID) investigators. N. Engl. J. Med. **1997**, *337* (22), 1576–1583.
51. Nattel, S.; Talajic, M. Recent advances in understanding the pharmacology of amiodarone. Drugs **1988**, *36* (2), 121–131.
52. Woosley, R.L.; Chen, Y.; Freiman, J.P.; Gillis, R.A. Mechanism of the cardiotoxic actions of terfenadine. J. Am. Med. Assoc. **1993**, *269* (12), 1532–1536.
53. Pratt, C.M.; Mason, J.; Russell, T.; Reynolds, R.; Ahlbrandt, R. Cardiovascular safety of fexofenadine HCl. Am. J. Cardiol. **1999**, *83* (10), 1451–1454.
54. Hoffmeister, H.M.; Beyer, M.E.; Seipel, L. Hemodynamic effects of antiarrhythmic compounds: intrinsic effects and autonomic modulation. Am. J. Cardiol. **1997**, *80* (8A), 24G–30G.

55. Mounsey, J.P.; DiMarco, J.P. Cardiovascular drugs: dofetilide. Circualtion **2000**, *102* (21), 2665–2670.
56. Hohnloser, S.H.; Singh, B.N. Proarrhythmia with class III antiarrhythmic drugs: definition, electrophysiologic mechanisms, incidence, predisposing factors, and clinical implications. J. Cardiovasc. Electrophysiol. **1995**, *6* (10 Pt 2), 920–936.
57. Zipes, D.P.; Jalife, J., Eds. *Cardiac Electrophysiology: from Cell to Bedside*; WB Saunders: Philadelphia, 2000.
58. Antzelevitch, C.; Sicouri, S. Clinical relevance of cardiac arrhythmias generated by afterdepolarizations: role of M cells in the generation of U waves, triggered activity and torsade de pointes. J. Am. Coll. Cardiol. **1994**, *23* (1), 259–277.
59. el-Sherif, N.; Caref, E.B.; Yin, H.; Restivo, M. The electrophysiological mechanism of ventricular arrhythmias in the long QT syndrome: tridimensional mapping of activation and recovery patterns. Circ. Res. **1996**, *79* (3), 474–492.
60. Antzelevitch, C.; Shimizu, W.; Yan, G.X.; Sicouri, S. Cellular basis for QT dispersion. J. Electrocardiol. **1998**, 30 Suppl., 168–175.
61. Vos, M.A.; Verduyn, S.C.; Gorgels, A.P.; Lipcsei, G.C.; Wellens, H.J. Reproducible induction of early afterdepolarizations and torsade de pointes arrhythmias by *d*-sotalol and pacing in dogs with chronic atrioventricular block. Circulation **1995**, *91* (3), 864–872.
62. Verduyn, S.C.; Vos, M.A.; van der Zande, J.; Kulcsar, A.; Wellens, H.J. Further observations to elucidate the role of interventricular dispersion of repolarization and early afterdepolarizations in the genesis of acquired torsade de pointes arrhythmias: a comparison between almokalant and *d*-sotalol using the dog as its own control. J. Am. Coll. Cardiol. **1997**, *30* (6), 1575–1584.
63. Gillis, A.M. Effects of antiarrhythmic drugs on QT interval dispersion—relationship to antiarrhythmic action and proarrhythmia. Prog. Cardiovasc. Dis. **2000**, *42* (5), 385–396.
64. Stambler, B.S.; Wood, M.A.; Ellenbogen, K.A.; Perry, K.T.; Wakefield, L.K.; VanderLugt, J.T. Efficacy and safety of repeated intravenous doses of ibutilide for rapid conversion of atrial flutter or fibrillation. Ibutilide repeat dose study investigators. Circulation **1996**, *94* (7), 1613–1621.
65. Lehmann, M.H.; Hardy, S.; Archibald, D.; Quart, B.; MacNeil, D.J. Sex difference in risk of torsade de pointes with *d,l*-sotalol. Circulation **1996**, *94* (10), 2535–2541.
66. Hohnloser, S.H. Proarrhythmia with class III antiarrhythmic drugs: types, risks, and management. Am. J. Cardiol. **1997**, *80* (8A), 82G–89G.
67. Torp-Pedersen, C.; Moller, M.; Bloch-Thomsen, P.E.; Kober, L.; Sandoe, E.; Egstrup, K.; Agner, E.; Carlsen, J.; Videbaek, J.; Marchant, B.; Camm, A.J. Dofetilide in patients with congestive heart failure and left ventricular dysfunction. Danish investigations of arrhythmia and mortality on dofetilide study group. N. Engl. J. Med. **1999**, *341* (12), 857–865.
68. Middlekauff, H.R.; Stevenson, W.G.; Saxon, L.A.; Stevenson, L.W. Amiodarone and torsades de pointes in patients with advanced heart failure. Am. J. Cardiol. **1995**, *76* (7), 499–502.
69. Vitola, J.; Vukanovic, J.; Roden, D.M. Cisapride-induced torsades de pointes. J. Cardiovasc. Electrophysiol. **1998**, *9* (10), 1109–1113.

70. Choy, A.M.; Darbar, D.; Dell'Orto, S.; Roden, D.M. Exaggerated QT prolongation after cardioversion of atrial fibrillation. J. Am. Coll. Cardiol. **1999**, *34* (2), 396–401.

71. Crumb, W.J., Jr.; Wible, B.; Arnold, D.J.; Payne, J.P.; Brown, A.M. Blockade of multiple human cardiac potassium currents by the antihistamine terfenadine: possible mechanism for terfenadine-associated cardiotoxicity. Mol. Pharmacol. **1995**, *47* (1), 181–190.

72. Cavero, I.; Mestre, M.; Guillon, J.M.; Crumb, W. Drugs that prolong QT interval as an unwanted effect: assessing their likelihood of inducing hazardous cardiac dysrhythmias. Expert Opin. Pharmacother. **2000**, *1* (5), 947–973.

73. Liu, X.K.; Katchman, A.; Drici, M.D.; Ebert, S.N.; Ducic, I.; Morad, M.; Woosley, R.L. Gender difference in the cycle length-dependent QT and potassium currents in rabbits. J. Pharmacol. Exp. Ther. **1998**, *285* (2), 672–679.

74. Jo, S.H.; Youm, J.B.; Lee, C.O.; Earm, Y.E.; Ho, W.K. Blockade of the HERG human cardiac K(+) channel by the antidepressant drug amitriptyline. Br. J. Pharmacol. **2000**, *129* (7), 1474–1480.

75. Kang, J.; Wang, L.; Cai, F.; Rampe, D. High affinity blockade of the HERG cardiac K(+) channel by the neuroleptic pimozide. Eur. J. Pharmacol. **2000**, *392* (3), 137–140.

76. Zhou, Z.; Vorperian, V.R.; Gong, Q.; Zhang, S.; January, C.T. Block of HERG potassium channels by the antihistamine astemizole and its metabolites desmethylastemizole and norastemizole. J. Cardiovasc. Electrophysiol. **1999**, *10* (6), 836–843.

77. Yap, Y.G.; Camm, A.J. Arrhythmogenic mechanisms of non-sedating antihistamines. Clin. Exp. Allergy **1999**, *29* (Suppl. 3), 174–181.

78. Anyukhovsky, E.P.; Sosunov, E.A.; Gainullin, R.Z.; Rosen, M.R. The controversial M cell. J. Cardiovasc. Electrophysiol. **1999**, *10* (2), 244–260.

79. Eckardt, L.; Haverkamp, W.; Borggrefe, M.; Breithardt, G. Experimental models of torsade de pointes. Cardiovasc. Res. **1998**, *39* (1), 178–193.

80. Carlsson, L.; Almgren, O.; Duker, G. QTU-prolongation and torsades de pointes induced by putative class III antiarrhythmic agents in the rabbit: etiology and interventions. J. Cardiovasc. Pharmacol. **1990**, *16* (2), 276–285.

81. Groh, W.J.; Gibson, K.J.; McAnulty, J.H.; Maylie, J.G. Beta-adrenergic blocking property of *dl*-sotalol maintains class III efficacy in guinea pig ventricular muscle after isoproterenol. Circulation **1995**, *91* (2), 262–264.

82. Morganroth, J.; Brown, A.M.; Critz, S.; Crumb, W.J.; Kunze, D.L.; Lacerda, A.E.; Lopez, H. Variability of the QTc interval: impact on defining drug effect and low-frequency cardiac event. Am. J. Cardiol. **1993**, *72*, 26B–31B.

83. Molnar, J.; Zhang, F.; Weiss, J.; Ehlert, F.A.; Rosenthal, J.E. Diurnal pattern of QTc interval: how long is prolonged? Possible relation to circadian triggers of cardiovascular events. J. Am. Coll. Cardiol. **1996**, *27*, 76–83.

5

Development and Application of Interspecies Biomarkers in Nonclinical Safety Evaluations

Frank D. Sistare
Food and Drug Administration, Laurel, Maryland, U.S.A.

I. INTRODUCTION

By the time a drug candidate has evolved through the developmental pipeline to submission for regulatory approval to begin phase I clinical trials, an average of 999 other candidates have been discarded. And the odds of that selected drug ultimately receiving marketing approval are estimated at approximately one in five [1]. A 20% clinical drug development success rate from that point suggests that the interface created by drug development approaches and regulatory review requirements being used over the past decades are in need of great improvement. Unacceptable toxicities and safety concerns are cited to account for 20–40% of the failures [2]. One factor for this may be that many of the assumptions implicit in current practice may be flawed or that the data generated are insufficiently informative. Once a compound has been selected to advance to the stage of initiation of clinical trials, its fate has essentially been cast. The clinical stages of investigation remain primarily to match the patient population to the indication, establish optimal dosing and use, and demonstrate proof of safety and efficacy for that selected molecule. The selection of the molecule is based on all of the study

This chapter was written in a private capacity with no official support by the Food and Drug Administration intended or implied.

information preceding clinical trial initiation—discovery data, mechanistic studies, chemistry, biology, in vitro toxicology, and the animal toxicology study findings.

Once the information from all of these disciplines has been integrated, the essential challenges of nonclinical studies preceding the clinical phases of development are to select the compound with greatest chance of clinical success, choose the most appropriate animal and nonanimal models to assess potential for toxicity to humans, determine the dose-limiting toxicities for that candidate, predict whether the toxicity dose-response profile will allow the achievement of doses that will produce efficacy in humans, and establish the starting dose for initiation of clinical trials. The assumptions are that the nonanimal models or strain/breed (often homogeneous inbred and healthy) and species of animal model chosen will substitute as accurate surrogate predictors for the wide spectrum of human diversity in unhealthy targeted patient populations. Humans can possess polymorphisms in both intended and unintended pharmacodynamic target molecules, as well as in molecules responsible for pharmacokinetic distribution and disposition. Humans do not all eat the same defined standard diet, drink only purified water, avoid simultaneous use of other medications, or live under the same pathogen-free environmental conditions. It is likely, therefore, that not only would differences between humans and animal studies be expected, but also wider differences in responses among humans are to be expected. Critical decisions must be made using the best available preclinical data to judge the impact of those expected, but difficult-to-predict, differences on clinical drug development success.

II. CURRENT PRACTICE: USE OF BIOMARKERS IN MEETING THE OBJECTIVES OF NONCLINCAL STUDIES

Data from animal toxicology and safety pharmacology studies are needed from drug developers to demonstrate for regulatory agencies a safe strategy for initiating human clinical trials. A thorough assessment from nonclinical studies should (1) identify the dose-limiting and associated toxicities and whether the toxicities seen are reversible, (2) identify safety pharmacology risks of the drug (and its metabolites) to function of the central nervous system, cardiovascular system, gastrointestinal system, immune system, pulmonary system, and other systems, (3) define the toxicokinetics, i.e., dose/exposure/time relationship, (4) define the safe starting dose for human trials, and (5) define an appropriately safe patient-monitoring strategy for clinical trials. Animal toxicology studies are terminated with a necropsy and microscopic examination of tissue specimens to identify evidence for drug-induced pathology. In animal studies there is high dependence on histopathological evaluations of tissues to delineate the toxicological profile of pharmaceutical candidates. Gross and microscopically visible alterations in cellular and tissue morphology, staining differences, alterations in compartmental tissue integrity, alterations in cellular integrity, host defense cell infiltrations, and other factors are interpreted in the context of mechanistic understanding of drug action, understanding of species differences and similarities, and measures of drug exposure levels for interpreting likely significance of animal toxicology study findings to humans.

Serum concentrations of drug and metabolites have become highly useful biomarkers of exposure. The therapeutic ratio for an agent is defined as the multiple between the exposure achieved at the highest dose (NOAEL, or no observable adverse effect level) at which no toxicity is discovered in a given species, as compared to the exposures achieved at the highest doses targeted for achieving evidence for efficacy in that species. Since proof of drug efficacy in nonclinical animal studies is not commonly a regulatory mandate, the focus of nonclinical studies is almost exclusively on defining toxicity parameters for guiding a safe clinical development strategy. The therapeutic ratio is very useful, but underutilized, algorithm for benchmarking nonclinical study pharmaco-kinetic–pharmacodynamic relationships [3,4] in an attempt to estimate clinical performance. Another useful and far more commonly utilized measure for estimating clinical performance and also for guiding clinical dose selection is the safety margin. The safety margin is defined as the multiple between the exposure achieved at the highest dose at which no dose-limiting toxicity is discovered in the most sensitive animal species, as compared to the exposure achieved (or expected) at the highest dose targeted for achieving clinical efficacy. If the safety margin is sufficiently broad such that expected benefit outweighs likely risk, then expectations are high that toxicity limitations will not affect successful clinical trial outcome. Important to this assessment is a comparison of the metabolite profile between the test species and humans, which must often be made without direct knowledge of the biological activities of the various individual metabolites. In current regulatory toxicology practice, emphasis is heavy on measures of circulating drug molecule and metabolite concentrations as biomarkers of exposure that will be used as common "interspecies" reference points to assess relevancy of animal toxicology study findings and help to guide clinical development plans. Often missing from these analyses, however, are similarly extrapolatable interspecies measures of toxicodynamic response.

There are animal study endpoints, however, that serve more than to identify clinical monitoring concerns and guide subsequent clinical trial conduct. These are animal study findings that are not monitorable in clinical trials and for which preclinical findings serve as the most definitive information for predicting human health effects in the absence of an ability to collect clinical data to directly address the question. For these datasets the animal study serves as the surrogate or replacement in lieu of definitive human data. The label that accompanies drug-marketing approval will cite the results of such animal studies. Conventional preclinical studies that serve as replacement surrogates for human studies include lifetime carcinogenicity, genotoxicity, photocarcinogenicity, and reproductive and developmental toxicity testing batteries (Table 1). Furthermore, when animal toxicities are encountered that cannot be safely monitored in the clinic, reliance is high on safety margin and risk-benefit judgments, and the animal study findings may essentially serve as human surrogate, with the drug label reflecting animal study findings if marketing approval is granted.

Experience, for example, with the well-documented animal and human teratogenicity of thalidomide and retinoids indicates that these animal models can serve as accurate predictors for such human toxicities. Such animal study findings are deemed to be reasonably predictive for these human toxicities and are therefore considered as surrogate models for identifying human reproductive risk potential.

For evaluation of lifetime human carcinogenicity risk, the linkage of human tumor endpoints to drug exposures is only rarely reliably confirmed, refuted, or evaluated in humans because of ethical, scientific, and practical reasons. The strength of the human data exists primarily through epidemiology studies and rarely from prospective clinical trial designs. Therefore, there is high reliance on the interpretation of rodent carcinogenicity study findings as the best and most definitive information that can be reasonably obtained. Findings from such rodent carcinogenicity studies are incorporated into product labeling. There is no human carcinogen identified by epidemiological data that has not been demonstrated to be carcinogenic in animal studies, except some have raised arguments that possibly the human carcinogen arsenic may not be carcinogenic to

Table 1 Nonclinical Study Models for Which Clinical Trial Data Generally Do Not Substitute

Toxicity	Model system
Genotoxicity	Bacterial mutation assays
	Eukaryotic cell culture systems
	Mouse/rat/dog in vivo micronucleus assays
Reproductive and development	Rodent/rabbit teratology studies
Carcinogenicity	Rodent 2-year lifetime bioassays
	Short- and intermediate-term alternative rodent bioassays
Photocarcinogenicity	SKH1 hairless mouse 1-year photocarcinogenicity assay
Host resistance	Tumor, viral, fungal, parasite, etc., challenged mouse

rodents [5]. There is evidence, however, demonstrating that many compounds found to be carcinogenic in rodent bioassays are likely not to be human carcinogens [6]. For this reason, much effort is placed on understanding and interpretation of rodent carcinogenicity study findings, and more recently on evaluating alternative transgenic models for improving the science of carcinogenicity testing [7]. In this regard, evaluations of the human relevancy of certain animal tumor findings have been investigated through well-designed follow-up mechanistic animal studies incorporating biomarker analyses (Table 2). In such cases, dose-dependent tumorigenesis can be linked to the same dose-dependent alteration in mechanistically linked biomarker alterations, and intervention of biomarker-dependent mechanisms can often be demonstrated to block tumor induction. Findings may be judged to be of little relevance to humans if the mechanism is specific for rodents, or if the doses associated with biomarker-dependent mechanisms will not cause the same alterations in the clinic. Knowledge of mechanism, together with a knowledge of comparative biochemistry between rodents and humans, allows a logical perspective to judge relevance for human risk. For example, when induction of rodent thyroid follicular cell tumors has been observed in 2-year bioassays, monitoring of

Table 2 Rodent Tumor Findings for Which Human Relevancy Concerns Have Been Reduced Using Comparative Interspecies Biomarker Measurements

Rodent tumor	Mechanism	Biomarker	Ref.
Mammary gland neoplasms	Dopamine antagonist (e.g., haloperidol) stimulation of prolactin secretion by reducing dopamine inhibition of pituitary	Serum prolactin elevation	8
Thyroid gland follicular cell tumors	Pituitary–thyroid axis derangement: induced metabolism of T3/T4, interference of T3/T4 synthesis, reduction of high-affinity thyroglobulin	Serum TSH elevation as a function of serum T3/T4; increased urinary T3/T4 turnover	9
Leydig cell tumors	Dopamine agonist (e.g., mesulergine) inhibition of pituitary prolactin secretion	Prolactin decrease; serum LH increase	10
Gastric enterochromaffin cell–like carcinoid tumors	Prolonged inhibition of gastric acid secretion stimulates increase in gastrin levels	Serum gastrin increase	11

thyroid gland activity by measuring T3/T4 turnover and serum levels of TSH as a function of dose has been used to demonstrate critical differences between human and rodents. The results can account for the appearance of rodent thyroid tumors with sulfonamides and phenobarbital, for example, and have reduced concern for human thyroid tumorigenesis with these compounds.

Chronic inhibition of pituitary prolactin secretions in a 2-year female rat cancer bioassay by dopamine agonists such as bromocryptine results in a measurable endocrine imbalance that has been associated with endometrial tumor development. In male rats, Leydig cell tumors may result when chronic prolactin inhibition may similarly result in endocrine imbalance, that is dependent upon pituitary LH secretion. Chronic stimulation of pituitary prolactin secretion in the female rat in response to sustained dopamine blockade or depletion results in chronic proliferative stimulation of mammary gland tissue, that can progress to neoplasia. Sustained inhibition of gastric acid secretion by protein pump inhibitors or by histamine type 2 receptor antagonists can result at high doses in a sustained significant elevation of gastrin, which can stimulate gastric ECL-like tumor development.

Understanding of mechanism and a comparative analysis of dose-biomarker response between rodents and humans are critical to assessing risk relevance at human exposure levels. It should be noted that it is inappropriate to assume that all rodent tumor findings at the sites described in Table 2 are irrelevant to humans unless a similar hormonal imbalance feedback loop relationship can be ruled in and human relevance at human exposure levels has been investigated.

Animal studies are routinely applied to define the relationship between dose, drug exposure, and dose-limiting toxicity to evaluate appropriateness for clinical trials. In such animal studies involving extensive histopathological evaluations, the use of serum or plasma biomarkers is often limited, in routine analyses, as described by a working group of clinical pathologists in 1996 [12]. This working group has recommended routine application of a core set of clinical chemistry endpoints as listed in Table 3 for routine toxicity screening. These routine measures together with hematology data serve as reporters of diminished organ functions (e.g., BUN, creatinine), reporters of altered tissue integrity (e.g., ALT, AST), reporters of altered general homeostasis (e.g., electrolytes, pH), or reporters of tissue response to injury (e.g., alterations in circulating cell populations). Together with the results from histopathology from the same toxicology studies, and with data from safety pharmacology studies, a strategic plan for monitoring safety in clinical trials is developed.

These same clinical pathology biomarkers have been used for routine monitoring for well over 25 years despite great strides in biological research over this same time frame. The integrity of most tissues is not being monitored using the above core set of clinical pathology biomarkers. Many of the clinical

Table 3 Animal Clinical Pathology 1996 Routine Testing Recommendations[a]

Hematology	Clinical chemistry
Total leukocyte count	Glucose
Absolute differential leukocyte count	Urea nitrogen
Erythrocyte count	Creatinine
RBC morphology evaluation	Total protein
Platelet count	Albumin
Hemoglobin concentration	Calculated globulin
Hematocrit	Calcium
Mean corpuscular volume	Sodium
Mean corpuscular hemoglobin	Potassium
Mean corpuscular hemoglobin concentration	Total cholesterol
Hemostasis	Hepatocellular: 2 of the following
Prothrombin time	Alanine aminotransferase
Activated partial thromboplastin time	Aspartate aminotransferase
Platelet count	Sorbitol dehydrogenase
Urinalysis	Glutamate dehydrogenase
Color and turbidity	Total bile acids
Volume	Hepatobiliary: 2 of the following
Specific gravity or osmolality	Alkaline phosphatase
pH	Gamma glutamyl transferase
Total protein	5'-nucleotidase
Glucose	Total bilirubin
	Total bile acids

[a] Cited from Ref. 12.

pathology biomarkers have high background serum concentrations normally and derive from multiple tissues. These characteristics account for both reduced biomarker sensitivity and reduced biomarker specificity. While histopathology can be seen in association with doses that can illicit clinical chemistry alterations, tissue histopathology is frequently seen at times and doses prior to elevation in any of these clinical pathology serum biomarkers. It can be argued, therefore, that through careful histopathological analyses, dose-limiting toxicities will be identified from nonclinical animal toxicology studies. However, the lack of a practical accessible biomarker-monitoring strategy can leave the sponsor and regulator to sort through speculative discussions surrounding relevance of animal findings to humans, and the inaccuracies of any defined exposure–response relationships during the proposed ensuing clinical phases of development. There is hope and great promise in the capability of genomic, proteomic, and metabonomic technologies to assist with questions of interspecies relevance by helping to define pathogenic mechanisms and by providing additional monitorable interspecies biomarkers.

III. CURRENT NEEDS

The mechanistic etiologies, clinical relevance, and clinical significance of drug-induced lesions in animal studies are rarely well understood initially. Additional studies may be needed to focus on (1) resolving uncertainties by increasing understanding of potential pathogenic mechanisms, and (2) identifying accessible interspecies biomarkers to diagnose and monitor for the early onset of dose-limiting toxicities.

There is a present need to consider expanding and improving upon the set of clinical pathology biomarker endpoints. For example, there is no routine recommendation for evaluation of drug-induced cardiac or skeletal muscle injury. Creatine kinase is not, for example, recommended as a routine biomarker. Such an additional test may be useful if a test material is suspected of affecting a specific tissue. Many additional biomarkers are now available at present commercially as specialty "esoteric" tests to address specific mechanistic questions relating to biomarker monitoring potential. Table 4 lists some of these. The validity of some of these tests for rat or dog is not well established at present and this may preclude greater utilization of measures of such endpoints. There may be much benefit to validating additional assays and developing high-throughput/low-cost tools for some of these analytes along with other carefully selected clinical chemistry markers that may be incorporated more routinely or systematically into rat and dog studies to improve evaluations of the linkage of monitorable parameters to histopathological outcomes.

Further research is needed in hypothesis-driven or discovery-based approaches to balance the cost versus the potential added value of expanding upon the set of routinely applied available interspecies biomarkers and incorporating such measures into early-evaluation paradigms. Focus is needed on establishing confidence in the linkage of clinical pathology alterations that may have been identified as sensitive indicators of animal morbidity to similar consequences in humans. It may be time to incorporate higher numbers of analytes into technological platforms with multiplexing capability, and to reevaluate a list of analytes to be routinely measured to enhance our ability to detect early onset of drug-induced injuries that can then be monitored similarly in the clinic. Alternatively, a tiered approach may be needed to incorporate more critically defined sets of mechanistic biomarkers for applications to specific questions following from initial routine protocols.

Strategies involving reliance on biopsies and histopathological evaluation for detection of toxicity are not attractive options for human studies. For human

Table 4 Additional Tests Available Through Commercial Clinical Veterinary Pathology Service Providers

Endocrine response	Organ/tissue integrity enzymes
ACTH	Aldolase
Aldosterone	Amylase
Androstenedione	Cardiac troponin I/T
Angiotensin-converting enzyme	CK isoenzymes
Angiotensin II	Isocitrate dehydrogenase
Antidiuretic hormone	Lipase
Atrial naturetic peptide	Intermediary metabolites
Calcitonin	2,3-diphosphoglycerate
Cholecystokinin	Free fatty acids
Cortiocosterone	Glycogen
Dihydrotestosterone	Glycohemoglobin AIC
Estradiol	HDL
FSH	Homocysteine
Gastrin	Lactate
Glucagon Growth hormone	LDL
IGF-1	Methemoglobin
Insulin	Orotic acid
LH	Pyruvate
Neuropeptide Y	Liver secreted
Parathyroid hormone	Alpha-1 acid glycoprotein
Progesterone	Alpha-2 macroglobulin
Prolactin	Apolipoproteins
Renin	Complement components
Secretin	C-reactive protein
Somatostatin	Serum amyloid A
Testosterone	Immune system
T3/T4	Anti-nuclear antibody
TSH	Histamine
	Immunoglobulins (A, G, E, M)

trials designed to evaluate dose-limiting clinical toxicities, the early detection of any drug-induced tissue injury would be greatly enhanced if well-characterized relevant and readily monitorable biomarkers were available to report organ injury without the need for visual inspection through biopsy or invasive or noninvasive imaging. And the concordance of those biomarkers with early onset and progression, not simply with the profound irreversible pathology seen at maximal doses and at study termination, could be established in animal studies. Potential substances may appear in the blood, urine, or other easily accessible body fluid that could be useful for diagnosing the early onset of drug-induced injury. The identification of biomarkers may be critical to developing a safe and successful clinical drug development plan, especially when the safety margin is expected not to be very broad, or the therapeutic index can be determined from animal studies to not be very broad. Potential biomarkers could include a range of substances that are linked to either the pathogenesis of the lesion or an early homeostatic response to the lesion, prior to irreversible tissue injury. Examples of such biomarkers may be proteins lost as integrity of specific cellular compartments is compromised; altered concentrations of carbohydrate, lipid, and protein metabolic intermediates; enhanced expression, secretion, or shedding of peptides, proteins, or surface molecules; or components of an inflammatory or immune response. Ideally, such endpoints would: (1) be very high in the tissue of interest but very low in normal sera, (2) be reliably and easily measurable, (3) be easily accessible and monitorable across species, (4) appear quickly and persist sufficiently long after initial elevation, and (5) be present only in the tissue of interest and therefore be specifically diagnostic for the expected injury.

As discussed earlier, measures of drug and metabolite serum and tissue concentrations serve as valuable biomarkers of exposure—for the purpose of relating blood levels at which effects are seen across various test species and that can then serve as a reference benchmark for moving into the clinic. Understanding the relationship between dose, blood level, and observed toxicity in animal studies is important for establishing the apparent safety margin for regulatory and drug development decision making in moving forward to human clinical trials. If, however, the only evidence of dose-limiting toxicity is histopathology and no readily accessible biomarker of the early onset of toxicity can be established in animal studies, reliance upon the monitoring of drug blood levels alone in human studies (Table 5), remains speculative. Without a scientifically sound reference response endpoint that cuts across species for linking an accessible biomarker alteration to the earliest elements of toxic effect or response, the monitoring of drug blood levels alone is lacking. Furthermore, given the set of low-sensitivity and low-specificity clinical pathology endpoints routinely incorporated in animal studies, confidence is high that improved biomarkers do exist and our abilities to sensitively detect the earliest aspects of tissue injury will improve.

In terminal animal toxicology studies, histological evidence of renal damage can often be seen at doses and at times that precede measurable rises in serum creatinine and urea nitrogen. It has been estimated that 50–70% of renal function can be compromised before a rise in serum creatinine and BUN would be expected [13]. In the clinic, numerous lifesaving oncological agents are associated with nephrotoxicity [14] and therefore should be used only with caution in patients at risk for renal dysfunction. The commonly used measure of serum creatinine has been observed to remain within the normal range in patients with cisplatin-associated decreases in renal function using the more stringent measure of ^{51}Cr–EDTA clearance [15,16], which is generally reserved for clinical research. Potentially more informative biomarkers of drug-induced kidney injury have been proposed. For example, the appearance of specific glutathione S-transferase isoenzymes in the urine following drug-induced renal injury may report early renal injury better than serum creatinine and that may distinguish damage to proximal tubule (appearance of alpha-GST) from damage to distal tubule (appearance of pi-GST) [17–20]. Urinary cytokines have been shown for example to be useful markers of reflux nephropathy. Evidence suggests that urinary cytokine measurements may serve as improved markers of renal dysfunction over less sensitive markers such as serum creatinine [21,22]. Utility in drug-induced damage is unknown.

In terminal animal toxicology studies, histopathological evidence of myocardial damage can be seen without associated increases in any of the biomarkers cited in Table 5. Even when creatine kinase MB isoenzymes is measured, histopathology can be observed in rats treated with relatively high doses of isoproterenol without serum CK-MB alterations that is, however, associated with measurable increases in serum levels of cardiac troponin T [23]. Similarly, with weekly administration of doxorubicin to rats, serum cardiac troponin T elevations correlate with the cumulative doses associated with onset of myofibrillar loss, vacuolization, and myocardial degeneration [24,25]. For diagnosing clinical acute myocardial infarction, a consensus recommendation has been made to treat any reliably detected serum elevation of a cardiac troponin as abnormal and is preferred over CK-MB [26]. In the clinic the endomyocardial biopsy has been used to definitively document drug-induced cardiac disease [27] but is largely reserved for clinical research and not a practical sample for bedside monitoring. Small rises in cardiac troponin T concentrations have shown initial promise for predicting risk for myocardial injury in pediatric patients being

Table 5 Examples of Drug-Induced Toxicities Encountered in Animal Studies with No or Poorly Monitorable Biomarkers

---

Phospholipidosis

Kidney damage—e.g., vacuolization, glomerular nephropathy (BUN/Cr are late-appearing)

Induction phase of drug allergy

Vasculitis

Neurotoxicities—e.g., neuronal vacuolization, demyelination

Focal organ necrosis/fibrosis

Arthropathies—seen in juvenile animal studies after destruction of chondrocytes and erosion of joint cartilage; irreversible once functional disturbance is visually apparent

Myocardial tissue damage

Gastrointestinal mucosal damage

---

treated with doxorubicin [28]. Other clinicians have similarly shown the added value of cardiac troponin serum measurements for therapeutic monitoring of drug-induced myocardial injury [29,30]. Incorporation of such "nonroutine" clinical chemistry endpoints by drug sponsors into routine animal toxicology studies and even very focused mechanistic studies of investigational new drugs is lacking. Therefore, beginning during the clinical phases of development to query whether such biomarkers truly possess interspecies bridging capability will be difficult. Developing confidence and defining strengths and limitations of such biomarkers may best be initiated during controlled diverse tests of sensitivity and specificity in the earlier nonclinical phases of a drug development plan.

IV. THE PATH FORWARD: OPPORTUNITIES USING STRATEGIES INVOLVING PRECLINICAL STUDY DESIGNS

This chapter has focused on the value discerned from measures of biomarkers of exposure and biomarkers of effect in animal studies. Biomarkers of effect for this discussion can be considered in the following context to signal: (1) an inconsequential correlative perturbation, (2) a diagnostic perturbation linked proximately to toxicity (diagnostic of compromised function or altered integrity), (3) a causative or mechanistically contributory perturbation that may be far upstream of actual toxicity, or (4) an evolutionarily conserved attempt by the host at a corrective homeostatic response following injury. An understanding of which role(s) the biomarker is fulfilling is important for integrating the weight of evidence that will be used to interpret and develop confidence in the value of the biomarker to safety guide the next phases of clinical development.

In contrast to the use of the surrogate models described above for which no human biomarker equivalent can be sought in a clinical trial, the biomarkers used to monitor for pathological consequences seen in animal toxicology studies are endpoints that could settle questions of concern and of relevance of animal toxicology study findings to humans. Unlike the surrogate animal models, these studies are not providing untestable *predictions* of human outcome, but rather testable hypotheses based on objectively measurable biomarkers that are expected to *report* any similar drug actions resulting in similar injury to humans. If the biomarkers are detectable and serve to report late toxicity, only when pathology is severe, progressive, and irreversible, little value is added. On the other hand, a biomarker that can report the very early onset of tissue injury may still be of great value if the test is sensitive, the damage is minimal, and the toxicity is reversible or if rapid intervention will halt progression and limit morbidity. There will clearly be a continuum of practical utility and ultimate application of these biomarkers whose limitations within this continuum could be explored, recognized, and established. Animal studies are valuable in this regard toward establishing where in this continuum of practical utility these biomarkers may lie. The mechanistic linkage between histopathology and measurable biomarker alterations can be more thoroughly investigated in animal studies. Reversibility of histopathology and biomarker alterations can similarly be explored carefully in controlled animal research studies. Such endpoints are highly supportive for allowing clinical testing to proceed and could provide very valuable collective experience toward further defining future utility of candidate toxicity biomarker endpoints as their value is further evaluated in the clinic.

Defining the utility of known but unevaluated endpoints as useful markers for marking drug-induced injury is one rationale for designing animal studies. In addition, the opportunities may be greatest for discovering new and improved biomarkers using animal models where toxicities can be induced intentionally, but humanely, under carefully controlled study conditions with definitive "gold standard" histological confirmation of extent of tissue injury. One major limitation has been the lack of generally available test reagents for proteins expressed in the rat and dog, which have served for decades as the main species for toxicity testing. Reagents to measure human and mouse proteins are generally more widely available with generally unknown rat or dog protein cross-reactivity.

Technological advances are beginning to address this with the capability for nucleic acid cloning and protein sequencing and the capability for preparing monoclonal and monoclonal-like polyclonal antibodies against specific sequence epitopes. But tools for accurate measurement of serum analytes in rats and dogs that may serve as useful biomarkers represent a major unfilled need. Inherent in the phrase "useful biomarkers" is the implicit assumption of the existence of proven robust and consistently useful analytical tools to measure those analytes.

Investigative approaches to identify such biomarkers could include, but are not limited to, technologies incorporating genomics, proteomics, and metabonomics, which are increasingly being applied to drug toxicity studies. Experiments using genomic-based microarray technologies directed at analyses of global expression profiling have demonstrated that cellular responses to drugs and toxicants can involve coordinately expressed members of biochemical pathways that enable grouping of agents according to mechanism of action [31–34]. Such use of genomics with in vitro cell lines and with affected target tissues from animal toxicology studies is expected to more quickly predict biological activity, to identify classes of toxicity that are of high concern, and to enhance understanding of mechanisms involved in the generation of histopathology. Starting with measurements of 1200 genes, Thomas et al.[34] demonstrated that the correct prediction for 24 compounds among five toxicological classes was 100% accurate with analysis focused only on 12 diagnostic gene transcripts. Such data suggest that applications of real-time quantitative polymerase chain reaction (RTQ-PCR) or similar technology to accurately and precisely measure expression of only a minimal number of carefully selected small sets of individual biomarker genes may become more generally applied over time.

These expression-based genomic technologies are expected to greatly accelerate the iterative process for identifying toxicophores that could be designed out of the chemical structure of a pharmaceutical candidate early in drug discovery and development. This should improve the inherent quality performance characteristics of pharmaceutical candidates before ever coming to regulatory agencies with data for preclinical evaluations to support applications for clinical development [35]. Nevertheless there will still be the need for better easily accessible biomarkers to bridge to the clinic and improve clinical monitoring strategies, provide assurances against drug-induced irreversible injuries, and better demonstrate safety in clinical trials. After drug-marketing approval is granted, the need will remain for improved approaches to minimize the risk and severity of serious adverse events in the early time period of exposure to the broader patient population. Genomic technologies may provide bridging biomarker endpoints applicable from animals to the clinic for especially vexing toxicity concerns if, for example, expression profiling of circulating peripheral blood leukocytes (PBL) can be shown to report the early onset of drug toxicities. Promising data indicate that unique PBL expression profiles are associated with seizures, stroke, hypoglycemia, and hypoxia [36]. As the linkage between gene expression pattern alterations in target organs is made to patterns of protein biomarker alterations appearing in sera, or in sentinel PBLs, even greater power should be derived from monitoring accessible interspecies biomarkers.

Proteomics technologies are similarly spurring the identification of protein toxicity biomarkers. One of the most successful approaches for separating proteins derived from biological matrices has been two-dimensional gel electrophoresis developed over 25 years ago [37]. Today, separated proteins are stained and imaged prior to mass spectrometric analysis. With the success in sequencing the human genome, the progress being made in sequencing genomes of several mammalian species, and the general conservation in sequence homology seen across mammalian species, genetic information is providing a template for the identification using mass spectrometry of specific proteins separated from biological matrices. Advances in sensitive protein stains, computer imaging, and mass spectrometry, together with the growing genomic and protein sequence databases, have spurred proteomic research [38]. Numerous examples for applications of this technology to define target tissue biomarkers of drug action and toxicity have been described; for example, see Refs. [39–42]. Two-dimensional gels can be limited in their ability to detect low-abundance proteins [43]. For discovery of drug-induced tissue-specific serum biomarkers, therefore, animal studies incorporating drug toxicities resulting in a spectrum of reversible to irreversible organ damage provide one potential solution to this limitation.

Alternatives to 2DGE-based technology include the use of two-dimensional microcapillary liquid chromatographic methods [44] and isotopically coded affinity tags (ICAT) to differentially label proteins from two separate populations to identify the proteins and to determine changes in relative abundance [45]. Surface-enhanced laser desorption ionization (SELDI) spectrometry has been applied to analyze protein from lysates of laser capture microdissected cellular subpopulations of tissues demonstrating the reproducible tissue proteomic biomarker profiles can be derived that are diagnostic of different tumor types and different stages of tumor progression [46]. Applications of SELDI using serum samples for early detection and diagnosis of human cancers is showing great promise [47]. We have shown recently that SELDI can similarly be used to discriminate sera from rats with isoproterenol-induced cardiac toxicity from normal controls [48]. The development of robust, sensitive protein arrays similar in design to gene expression cDNA arrays to simultaneously measure thousands of proteins has begun [49,50]. Another approach has been developed to measure scores of analytes in very small volumes of body biofluids using bead-coupled specific antibody reagents and fluorescent signal generation [51]. Such technology may have more immediate applicability to implementing approaches to questions depending upon more conventional assays using biomarkers that will be discovered by many of the methods described above.

Metabonomics technology, which relies on high-resolution ^{1}H-NMR spectra of urine and serum and multivariate statistical analysis to identify metabolic biomarker patterns, has been used to detect and understand metabolic responses to drug-induced toxicities [52]. Examples have included drug-induced renal toxicity [53], hepatotoxicity [54], and vasculitis [55]. Unlike gene expression microarrays and protein-antibody-based detection techniques, which rely on designing species-specific capture matrices, urinary metabonomics can simultaneously monitor hundreds of urinary trace biomolecules with the same machinery without the need to design species-specific capture molecules. Improved understanding of the specificity of the linkage between the urinary biomolecular changes with drug toxicities is important for this technology to realize its great promise for assisting with interspecies extrapolations and for monitoring toxicities. This technology draws power from the elucidation of an overall pattern of changes and employs a multivariate statistical approach to assist with comprehension. This trend in applying algorithms for assessing the contributions of hundreds and even thousands of measurable parameters to specify unique diagnostic patterns, none of which may individually supply sufficient predictive confidence, challenges the classic concept represented by current routinely measured biomarkers [47].

V. APPROACHES TO EVALUATING WHETHER THERE IS INCREASED VALUE FOR USE OF BIOMARKER MEASUREMENTS IN PRECLINICAL STUDIES

Further laboratory research is needed both to expand upon the available choices of biomarkers using technologies such as those described above, and to solidify the scientific evidence to objectively evaluate the utility of reliable biomarkers of drug-induced tissue injury to reduce risk and improve risk management. One productive route of biomarker evaluation has been to focus on those biomarkers that already have approved reagent kits for measuring clinical analytes in humans. When the challenge of deriving assay kit components that will cross-react with the rodent, dog, and monkey can be met, then the experiments can be performed to measure whether drug-induced injury will result in elevations of markers that are already measurable. The alternative is to discover the molecules that change and then develop the specific analytical tools. This is analogous to the serial analysis of gene expression or subtractive hybridization approaches to discover gene expression alterations linked to a perturbation of interest as compared to the expression microarray approach for which reagent targets preexist.

Approaches to ascertaining the value of knowledge of an analyte serum concentration in assessing drug-induced tissue injury might be phrased more generally at the outset. Will the knowledge of the serum concentration of an analyte add value by helping to identify individual patients that may be at increased risk for developing a progressive organ-directed disease? As mentioned earlier, certain oncological agents are associated with nephrotoxicity [14] and therefore should be used only with caution in patients at risk for renal dysfunction. Similarly, caution is advised in prescribing thiazolidinedione antidiabetic agents to patients with preexisting hepatic or congestive heart disease. Improved sets of sensitive markers for these and many other examples of underlying predispositions for drug-induced morbidity that can be routinely measured to better identify patients at risk and similarly be useful for monitoring for any early dose-related onset of such toxicity would be extremely valuable. Recent legislation promoting the expansion of clinical trials to include children adds even more impetus to the need for biomarkers that may serve a useful role in protecting against insidious long-term sequleae of drug-induced injury.

We know that humans are genetically diverse while we generally perform out toxicology studies in define inbred and fairly homogeneous animal strains. The reasons are readily apparent—the most prominent being the ability to minimize experimental variability and thereby enhance signal to noise using smaller numbers of animals. However, differences in susceptibilities to the toxic effects of chemical agents across strains of a given species can be prominent and have been well documented in the literature. In our exploration of the utility of biomarkers to separate pharmacological from toxicological actions, it may be prudent to incorporate experimental strategies involving strain differences to evaluate our ability to prospectively discriminate and identify animals at risk for developing toxicity.

Additional approaches to evaluating and establishing linkages for biomarker associations with drug-induced disease include the ability to include intervention approaches into animal studies to assess strength of linkage. If an intervention approach reduces both injury and biomarker levels, mechanistic association may not be proven but will add evidence to the strength of the association. With such strength of evidence, follow-up investigations may be directed at defining utility for monitoring progression of injury, for evaluating the success of alternative intervention approaches, and even possibly exploring whether the biomarker could, as well, be a target for intervention if it is on the causative pathway for injury.

VI. SUMMARY AND CONCLUSIONS

For initial animal toxicology studies, test compound is administered at doses that are intended to define the toxicities when maximally tolerated doses are exceeded. For decades, the components of clinical chemistry, hematology, and urinalyses that have been measured routinely as components of these studies have been minimal. The list of urine and serum components that are being identified is expanding, and improvements in technology are being made that will be able to identify deviations in these markers in rapid multiplexed assays. These two forces are challenging the long-held minimalist approach. One great challenge will be to demonstrate how much value that the added investment in measuring these markers in animal studies will add to improving clinical safety and improving drug development. Investigation of toxicokinetics, along with the time- and dose-dependent association between drug exposure and biomarker alterations, is certain to refine noninvasive monitoring capability. As more and better biomarkers become readily monitorable at the point of patient care, it is difficult to see that this information would not be valuable to pharmaceutical developers and regulators. The alternative is to resist change and ignore the challenges and opportunities that this new knowledge and these new technologies are presenting.

REFERENCES

1. http://www.allp.com/drug_dev.htm. Phases of Product Development.
2. Li, A.P. Screening for human ADME/tox drug properties in drug discovery. Drug Discovery Today **2001**, *6*, 357–366.
3. Levy, G. The case for preclinical pharmacodynamics. In *Integration of Pharmacokinetics. Pharmacodynamics, and Toxicokinetics in Rational Drug Development*; Yacobi, A., Skelly, J.P., Shah, V.P., Benet, L.Z., Eds.; Plenum: New York, 1993; 7–13.
4. Sheiner, L.B.; Steiner, J.L. Pharmacokinetic/pharmacodynamic modeling in drug development. Annu. Rev. Pharmacol. Toxicol. **2000**, *40*, 67–95.
5. Vahter, M. Methylation of inorganic arsenic in different mammalian species and population groups. Sci. Prog. **1999**, *82*, 69–88.
6. Monro, A. Are lifespan rodent carcinogenicity studies defensible for pharmaceutical agents? Exp. Toxicol. Pathol. **1996**, *48*, 55–166.
7. Robinson, D.E.; MacDonald, J.S. Background and framework for ILSI's collaborative evaluation program in alternative models for carcinogenicity assessment. Toxicol. Pathol. **2001**, *9* (S), 13–19.
8. O'Connor, J.C.; Plowchalk, D.R.; Van Pelt, C.S.; Davis, L.G.; Cook, J.C. Role of prolactin in chloro-S-triazine rat mammary tumorigenesis. Drug Chem. Toxicol. **2000**, *23*, 575–601.
9. Hood, A.; Hashmi, R.; Klaassen, C.D. Effects of microsomal enzyme inducers on thyroid–follicular cell proliferation, hyperplasia, and hypertrophy. Toxicol. Appl. Pharmacol. **1999**, *160*, 163–170.
10. Cook, J.C.; Klinefelter, G.R.; Hardisty, J.F.; Sharpe, R.M.; Foster, P.M.D. Rodent Leydig cell tumorigenesis: a review of the physiology, pathology, mechanisms, and relevance to humans. Crit. Rev. Toxicol. **1999**, *29*, 169–261.
11. Burek, J.D.; Patrick, D.H.; Gerson, R.J. Weight of biological evidence approach for assessing carcinogenicity. In *Carcinogenicity*; Grice, H.C., Cimina, J.L., Eds.; Springer-Verlag: New York, 1998; 83–95.
12. Weingand, K.; Brown, G.; Hall, R.; Davies, D.; Gossett, K.; et al. Harmonization of animal clinical pathology testing in toxicity and safety studies. Fundam. Appl. Toxicol. **1996**, *29*, 98–201.
13. Goldstein, R.S.; Schnellmann, R.G. Toxic responses of the kidney. In *Casarett and Doull's Toxicology: The Basic Science of Poisons*; Klaassen, C.D., Ed.; McGraw-Hill: New York, 1996; 417–442.
14. Kintzel, P.E. Anticancer drug-induced kidney disorders. Drug Safety **2001**, *24*, 19–38.
15. Daugaard, G.; Rossing, N.; Rorth, M. Effects of cisplatin on different measures of glomerular function in the human kidney with special emphasis on high-dose. Cancer Chemother. Pharmacol. **1998**, *21*, 163–167.
16. Womer, R.B.; Pritchard, J.; Barratt, T.M. Renal toxicity of cisplatin in children. J. Pediatr. **1985**, *106*, 659–663.

17. Harrison, D.J.; Kharbanda, R.; Cunningham, D.S.; McLellan, L.I.; Hayes, J.D. Distribution of glutathione S-transferase isoenzymes in human kidney: basis for possible markers of renal injury. J. Clin. Pathol. **1989**, *42*, 624–648.

18. Kharasch, E.D.; Horrman, G.M.; Thorning, D.; Hankins, D.C.; Kilty, C.G. Role of the renal cysteine conjugate B-lyase pathway in inhaled compound A nephrotoxicity in rats. Anesthesiology **1998**, *88*, 1624–1633.

19. Backman, L.; Appelkvist, E.L.; Ringden, O.; Dallner, G. Appearance of basic glutathion transferase in the urine during tubular complications in renal transplant recipients. Transplant. Proc. **1988**, *20*, 427–430.

20. Sundberg, A.; Appelkvist, E.L.; Dallner, G.; Nilsson, R. Glutathione transferases in the urine: sensitive methods for detection of kidney damage induced by nephrotoxic agents in humans. Environ. Health Perspect. **1994**, *102* (S), 293–296.

21. Gormley, S.M.C.; McBride, W.T.; Armstrong, M.A.; et al. Plasma and urinary cytokine homeostasis and renal dysfunction during cardiac surgery. Anesthesiology **2000**, *93*, 1210–1216.

22. Westhuyzen, J.; McGriffin, D.C.; McCarthy, J.; Fleming, S.J. Tubular nephrotoxicity after cardiac surgery utilising cardiopulmonary bypass. Clin. Chim. Acta **1994**, *228*, 123–132.

23. Bleuel, H.; Deschl, U.; Bertsch, T.; et al. Diagnostic efficiency of troponin T measurements in rats with experimental myocardial cell damage. Exp. Toxicol. Pathol. **1995**, *47*, 121–127.

24. Herman, E.H.; Zhang, J.; Lipshultz, S.E.; et al. Use of cardiac troponin T levels as an indicator of doxorubicin-induced cardiotoxicity. Cancer Res. **1998**, *58*, 195–197.

25. Herman, E.H.; Zhang, J.; Lipshultz, S.E.; et al. Correlation between serum levels of cardiac troponin-T and the severity of the chronic cardiomyopathy induced by doxorubicin. J. Clin. Oncol. **1999**, *17*, 2237–2243.

26. Albert, J.; Thygesen, K.; et al. Myocardial infarction redefined—a consensus document of the Joint European Society of Cardiology/American College of Cardiology Committee for the redefinition of myocardial infarction. J. Am. Coll. Cardiol. **2000**, *36*, 959–969.

27. Billingham, M.E. Role of endomyocardial biopsy in diagnosis and treatment of heart disease. In *Cardiovascular Pathology*, 2nd Ed.; Silver, M.D., Ed.; Churchill Livingstone: New York, 1991; 1465–1486.

28. Lipschultz, S. Ventricular dysfunction clinical research in infants, children and adolescents. Prog. Pediatr. Cardiol. **2000**, *12*, 1–28.

29. Cardinale, D.; Sandri, M.T.; Martinoni, A.; Tricca, A.; Civelli, M.; Lamantia, G.; Cinieri, S.; Martinelli, G.; Cipolla, C.M.; Fiorentini, C. Left ventricular dysfunction predicted by early troponin I release after high-dose chemotherapy. J. Am. Coll. Cardiol. **2000**, *36*, 517–522.

30. Morandi, P.; Ruffini, P.A.; Benvenuto, G.M.; LaVecchia, L.; Mezzena, G.; Raimondi, R. Serum cardiac troponin I levels and ECG/Echo monitoring in breast cancer patients undergoing high-dose ($7\,g/m^2$) cyclophosphamide. Bone Marrow Transplant. **2001**, *28*, 277–282.

31. Scherf, U.; Ross, D.T.; Waltham, M.; Smith, L.H.; Lee, J.K.; Tanabe, L.; Kohn, K.W.; Reinhold, W.C.; Myers, T.G.; Andrews, D.T.; Scudiero, D.A.; Eisen, M.B.;

Sausville, E.A.; Pommier, Y.; Botstein, D.; Weinstein, J.N. A gene expression database for the molecular pharmacology of cancer. Nat. Genet. **2000**, *24*, 236–244.

32. Burczynski, M.E.; McMillian, M.; Ciervo, J.; Li, L.; Parker, J.B.; Dunn, R.T., II; Hicken, S.; Farr, S.; Johnson, M.D. Toxicogenomics-based discrimination of toxic mechanism in HepG2 human hepatoma cells. Toxicol. Sci. **2000**, *58*, 399–415.

33. Waring, J.F.; Jolly, R.A.; Ciurlionis, R.; Lum, P.Y.; Praestgaard, J.T.; Morfitt, D.C.; Buratto, B.; Roberts, C.; Schadt, E.; Ulrich, R.G. Clustering of hepatotoxins based on mechanism of toxicity using gene expression profiles. Toxicol. Appl. Pharmacol. **2001**, *175*, 28–42.

34. Thomas, R.S.; Rank, D.R.; Penn, S.G.; Zastrow, G.M.; Hayes, K.R.; Pande, K.; Glover, E.; Silander, T.; Craven, M.W.; Reddy, J.K.; Jovanovich, S.B.; Bradfield, C.A. Identification of toxicologically predictive gene sets using cDNA microarrays. Mol. Pharmacol. **2001**, *60*, 1189–1194.

35. Ulrich, R.; Friend, S. Toxicogenomics and drug discovery: will new technologies help us produce better drugs? Nat. Rev. Drug Discovery **2002**, *1*, 84–88.

36. Tang, Y.; Lu, A.; Aronow, B.; Sharp, F. Blood genomic responses differ after stroke, seizures, hypoglycemia, and hypoxia: blood genomic fingerprints of disease. Ann. Neurol. **2001**, *50*, 699–707.

37. O'Farrell, P.H. High resolution two-dimensional electrophoresis of proteins. J. Biol. Chem. **1975**, *250*, 4007–4021.

38. Pandey, A.; Mann, M. Proteomics to study genes and genomes. Nature **2000**, *405*, 837–846.

39. Cunningham, M.L.; Pippin, L.L.; Anderson, N.L.; Wenk, M.L. The hepatocarcinogen methapyrilene but not the analog pyrilamine induces sustained hepatocellular replication and protein alterations in F344 rats in a 13-week feed study. Toxicol. Appl. Pharmacol. **1995**, *131*, 216–223.

40. Aicher, L. New insights into cyclosporine A nephrotoxicity by proteome analysis. Electrophoresis **1998**, *19*, 1998–2003.

41. Anderson, N.L. The effects of peroxisome proliferators on protein abundances in mouse liver. Toxicol. Appl. Pharmacol. **1996**, *137*, 75–89.

42. Steiner, S.; et al. Proteomics to display lovastatin-induced protein and pathway regulation in rat liver. Electrophoresis **2000**, *21*, 2129–2137.

43. Gygi, S.P.; Corthals, G.L.; Zhang, Y.; Rochon, Y.; Aebersold, R. Evaluation of two-dimensional gel electrophoresis-based proteome analysis technology. Proc. Natl. Acad. Sci. U.S.A. **2000**, *97*, 9390–9395.

44. Opiteck, G.J.; Lewis, K.C.; Jorgenson, J.W.; Anderegg, R.J. Comprehensive on-line LC/LC/MS of proteins. Anal. Chem. **1997**, *69*, 1518–1524.

45. Gygi, S.P.; Rist, B.; Gerber, S.; Turecek, F.; Gelb, M.H.; Aebersold, R. Quantiative analysis of complex protein mixtures using isotope-coded affinity tags. Nat. Biotechnol. **1999**, *17*, 994–999.

46. Pawaletz, C.P.; Gillespie, J.W.; Ornstein, D.K.; Simone, N.L.; Brown, M.R.; Cole, K.A.; Wang, Q.H.; Huang, J.; Hu, N.; Yip, T.T.; Rich, W.E.; Kohn, E.C.; Linehan, W.M.; Weber, T.; Taylor, P.E.; Emmert-Buck, M.R.; Liotta, L.A.; Petricoin, E.F., III. Biomarker profiling of stages of cancer progression directly from human tissue using a protein biochip. Drug Dev. Res. **2000**, *49*, 34–42.

47. Petricoin, E.F.; Ardekani, A.M.; Hitt, B.A.; Levine, P.J.; Fusaro, V.A.; Steinberg, S.M.; Mills, G.B.; Simone, C.; Fishman, D.A.; Kohn, E.C.; Liotta, L.A. Use of proteomic patterns in serum to identify ovarian cancer. Lancet **2002**, *359*, 572–577.

48. Ardekani, A.M.; Herman, E.H.; Sistare, F.D.; Liotta, L.A.; Petricoin, E.F., III. Molecular profiling of cancer and drug-induced toxicity using new proteomics technologies. Curr. Ther. Res. **2001**, *62*, 803–819.

49. Liotta, L.; Petricoin, E. Molecular profiling of human cancer. Nat. Rev. **2000**, *1*, 48–56.

50. Wettstein, D.A. Proteomics and protein chips. Clin. Lab. News **2001**, *July*, 22–24.

51. Carson, R.T.; Bignali, D.A.A. Simulatneous quantitation of 15 cytokines using a multiplexed flow cytometric assay. J. Immunol. Methods **1991**, *227*, 41–52.

52. Nicholson, J.K.; Lindon, J.C.; Holmes, E. "Metabonomics"—understanding the metabolic responses of living systems to pathophysiological stimuli via multivariate statistical analysis of biological NMR spectroscopic data. Xenobiotica **1999**, *29*, 1181–1189.

53. Holmes, E.; Nicholson, J.K.; Nicholls, A.W.; Lindon, J.C.; Connor, S.C.; Polley, S.; Connelly, J. The identification of novel biomarkers of renal toxicity using automatic data reduction techniques and PCA of proton NMR spectra of urine. Chemom. Intell. Lab. Syst. **1998**, *44*, 245–255.

54. Beckwith-Hall, B.M.; Nicholson, J.K.; Nicholls, A.W.; Foxal, P.J.D.; Lindon, J.C.; Connor, S.C.; Abdi, M.; Connelly, J.; Holmes, E. Nuclear magnetic resonance spectroscopic and principal components analysis investigations into biochemical effects of three hepatotoxins. Chem. Res. Toxicol. **1998**, *11*, 260–272.

55. Robertson, D.G.; Reily, M.D.; Albassam, M.; Dethloff, L.A. Metabonomic assessment of vasculitis in rats. Cardiovasc. Toxicol. **2001**, *1*, 7–19.

6

Validation of Assays for the Bioanalysis of Novel Biomarkers: Practical Recommendations for Clinical Investigation of New Drug Entities

Jean W. Lee
MDS Pharma Services, Lincoln, Nebraska, U.S.A.

Wendell C. Smith
Lilly Research Laboratories, Greenfield, Indiana, U.S.A.

Gerald D. Nordblom
Pfizer Global Research and Development, Ann Arbor, Michigan, U.S.A.

Ronald R. Bowsher
Lilly Research Laboratories, Indianapolis, Indiana, U.S.A.

I. INTRODUCTION

Biomarkers comprise a broad array of laboratory-based, physiological, and behavioral characteristics that are measured as indicators of normal biological processes, as pathogenic processes, or as pharmacological responses to therapeutic interventions [1]. With few exceptions, such as viral load, biomarkers are endogenous substances. Depending on the stage of drug development, biomarker measurements can provide important mechanistic, efficacy, or toxicity information. For example, biomarker data can be used to discover and select a lead compound, generate pharmacokinetic/pharmacodynamic (PK/PD) models [2,3] that aid in clinical trial design and expedite drug development, serve as surrogates for a clinical or mortality endpoint [4], and optimize drug therapy based on genotypic or phenotypic factors [5]. Additionally, biomarker data can be useful to define which patients to enroll in a drug study or to stratify patients within a clinical protocol.

A wide variety of in vitro analytical methodologies are used to quantify biomarkers, including chemical, colorimetric, chromatographic (e.g., liquid chromatographic–mass spectrometric, LC–MS–MS), immunochemical, and cell-based assays. Because lab-based biomarkers are so diverse, ranging from electrolytes to small molecules and macromolecules, it is not practical to address analytical validation in detail for all types of biomarker measurements. Therefore, we will focus our discussion on validation of in vitro assays for the bioanalysis of novel biomarkers with emphasis on quantitative binding assays to support the clinical investigation of new drug entities. In this chapter we have defined a novel biomarker as an analyte or activity that is measured by an in vitro assay that is not available as a routine clinical laboratory test. Alternatively, a novel biomarker could be an analyte or activity that is measured by "specialized analytical technology" that is not available in a routine clinical laboratory setting. Clearly, the analytical distinction between a novel and a routine biomarker (i.e., one that is quantified by means of an established clinical laboratory test) is subtle in some instances.

Biomarker assays are performed in a variety of laboratory settings, including pharmaceutical discovery laboratories, investigator sites, hospital laboratories, reference medical laboratories, and at contract research organizations (CROs). In contrast to routine clinical biomarkers, assays for novel biomarkers are often developed within a pharmaceutical company, biotechnology company, diagnostic company, or university research laboratory. For novel biomarker assays, the required level of analytical validation depends on the stage of drug development (Fig. 1). A strategy for parallel validation helps ensure consistency in the quality of analytical results between assays for the candidate drug and its novel biomarker. Clearly, rigorous bioanalytical method validation is usually not necessary for discovery-phase work, such as screening assays for novel drug targets. Because immunoassays and binding assays are prone to interference by the test sample matrix, the nature and design of the validation must change when biomarker assays are transitioned from a screening mode to one of bioanalysis, quantification of an analyte of interest in a biological matrix. Therefore, as the drug of interest progresses into preclinical and early-phase clinical evaluation (e.g., phase I), more thorough method validation is warranted. By the time the drug of interest has progressed into phase II studies full analytical validation is recommended, particularly if the biomarker assay is deemed to provide pivotal safety or efficacy information. We refer to this concept for progressive refinement and validation of novel biomarker assays as "stage-appropriate validation."

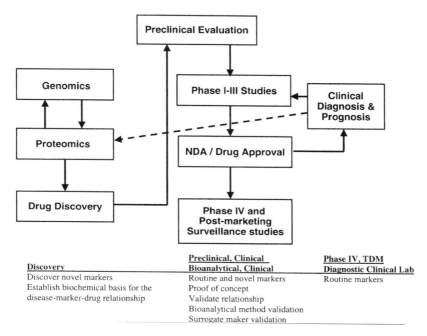

Figure 1   Application of biomarkers in various phases of drug development.

Even though the conference report for the 1990 Crystal City meeting on bioanalytical methods validation stated, "All pharmacodynamic measures for definitive bioequivalency or related studies must be fully validated under controlled conditions and should include a placebo"[6], limited information was included for in vitro biomarker assays. The new 2001 FDA guidance for bioanalytical method validation does not address validation of biomarker assays [7]. Consequently, in-depth guidance is lacking for validation of novel biomarker assays. In this chapter, we offer a strategy for validation of novel biomarker assays and stage-specific validation recommendations for clinical-phase studies.

II. CATEGORIZATION OF BIOMARKER ASSAYS

Unlike bioanalytical assays for conventional xenobiotic drugs in which quantitative results (i.e., plasma concentrations) are obtained by calibration against a highly purified and well-defined reference standard, biomarker assays can differ considerably depending on the type of analytical measurement and intended use of the reported result. Figure 2 summarizes the analytical

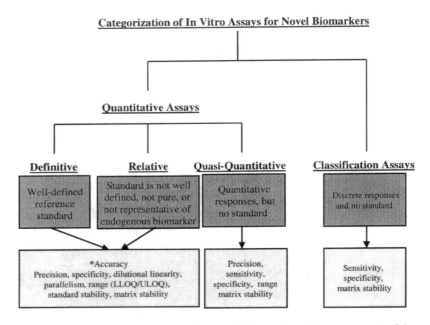

Figure 2   Categorization of assays for novel biomarkers. The upper set of boxes contains information about reference standard characteristics, while the lower set describes the analytical performance characteristics that are important for different categories of novel biomarker assays.

performance characteristics for different categories of biomarker measurements. For clinical-phase, laboratory-based biomarker assays, the preponderance of results is typically quantitative in nature. In the ideal situation, a "definitive" (absolute) quantitative measurement is obtained by calibration against a well-defined reference standard, which is representative of the endogenous biomarker. In the absence of a definitive standard, a "relative" result may be obtained by calibration against a reference material that is not well defined, not available in a purified form, or not fully representative of the endogenous biomarker. In the absence of a reference material, a quasi-quantitative* analytical result may be obtained by measurement of a test sample characteristic that is determined without comparison to a reference material and reported in continuous units of the test characteristic. Some examples of quasi-quantitative assays include measurement of enzymatic activity, determination of antidrug antibodies, or

---

*Definition of quasi-, having some resemblance by possession of certain attributes.

quantification of a vaccine antibody response in which the analytical response is reported as percent binding.

Unlike the previous "quantitative" categories, biomarker assays can in some instances produce qualitative analytical results that are discrete (discontinuous) and reported in terms of a characteristic of the test sample. A descriptive result (data type) may be either ordinal or nominal. Ordinal data are discrete numerical or character values that are spaced in a logical order, but a spacing interval is not implied (e.g., low, medium, and high, or $+$, $++$, and $+++$). Nominal data are discrete values with no implicit order that are used for the purpose of classification of results, such as yes/no or positive/negative. An example of a qualitative assay would be a method to detect the presence of a single nucleotide polymorphism or gene mutation in a sample of DNA.

For the purpose of categorization, we view assays of novel biomarkers as possessing characteristics of both clinical assays for the quantification of routine biomarkers and good laboratory practice (GLP)-compliant bioanalytical assays that are used for measuring drug concentrations. Even though validation recommendations for clinical assays (National Committee for Clinical Laboratory Standards, NCCLS) and GLP-compliant assays share similarities, differences do exist in their respective approaches. As described in Table 1, the major differences include the analytical performance characteristics evaluated during prestudy validation, the approaches used to assess accuracy, the procedure for establishing the lowest reportable concentration, and the specific criteria used for both method and run acceptance. These differences, coupled with the case-by-case nature required for validation of novel biomarker assays and the lack of formal guidance, have led to confusion among bioanalytical scientists about what procedures are necessary and appropriate for validating novel biomarker assays to support clinical investigation of new drug entities.

III.   ANALYTICAL ISSUES FOR NOVEL BIOMARKER ASSAYS

In contrast to bioanalytical assays for conventional xenobiotic drugs, the development and validation of assays for novel biomarkers are frequently accompanied by analytical issues that make adoption of a strict set of GLP-compliant validation rules impractical (Table 2). A number of these bioanalytical issues have been addressed in recent publications [8,9]. Particularly noteworthy for validation of quantitative assays for novel biomarkers are issues involving the reference standard. Ideally, the reference standard used for preparing the primary calibration standard should be homogeneous, pure, and identical to the endogenous analyte [10]. Pure standards and reference methods based on mass

Table 1 Comparison of Analytical Validation Approaches for Definitive Quantitative Assays

Analytical criteria	NCCLS[a] approach	GLP[b] approach
Analytical performance characteristics	Accuracy, precision, linearity,[c] specificity, assay sensitivity, recovery, method comparison, range, parallelism, stability of reagents.	Accuracy, precision, linearity,[c] specificity, LLOQ and ULOQ, recovery, assay sensitivity, parallelism, dilutional linearity, range, standard stability, analyte stability in biological matrix, stability of reagents.
Accuracy[d]	2 approaches are suggested: (1) assess bias after conducting a "split sample comparison expt.," in which the analyte is measured by the newly developed assay and another "gold standard" method, or (2) assess bias after using the new method to measure analyte concentration in an "official" or defined reference material.	Assessed by "spiking" experiments in which the drug of interest is quantified after its addition to the sample matrix at defined concentrations. Typically expressed as *mean bias* (%RE or %Recovery), which is the systematic difference between the mean measured concentration and the spiked nominal (theoretical true) concentration. *Note:* The 2001 FDA bioanalytical methods validation guidance recommends, when possible, that immunoassays be compared with a validated reference method using incurred samples to ensure lack of interference from substances that are similar physicochemically to the analyte.
Lowest reportable concentration	Generally, any response above the assay's sensitivity, "detection limit" (i.e., zero dose response ± 2 S.D.), is considered within the assay's reportable range.	Lower limit of quantitation, LLOQ, lowest analyte concentration in a sample that can be determined quantitatively with acceptable levels of accuracy and precision (i.e., 20% RE and 20% CV).

Prestudy validation (method acceptance criteria for accuracy and precision)	No specified minimal criteria for accuracy and precision; criteria depend on the intended clinical use.	±15% for RE and CV with 20% permissible at lower limit of quantitation. Recent publications have recommended the target values be made more lenient for biomarkers (e.g., ±25 and 30%). "4-6-X rule." Recommendations from the initial Crystal City workshop were 67% of QC results needed to be within ±20% of the nominal [6].
In-study validation (assay run acceptance)	Westgard rules (control charts) [25,26].	The new 2001 FDA guidance requires 67% of QC results to be within ±15% of the nominal [7]. In contrast, for immunoassays some recent publications have recommended making the run acceptance rule more liberal, particularly for macromolecules (e.g., proteins) [9,27].

[a] NCCLS documents: LAI-A2, Assessing the quality of radioimmunoassay systems—second edition: 14(17), 1994; I/LA9-T, A candidate reference method for serum digoxin: a model for radioimmunoassay reference methods, tentative guideline 16(1), 1996; I/LA21-P, Clinical evaluation of immunoassays; proposed guideline, 19(19), 1999.

[b] In addition to cited Refs. 6, 7, 9, 27, see Miller KJ, et al., Workshop on bioanalytical methods for macromolecules: summary report, Pharmac Res. 2001, 18(9), 1378–1383.

[c] Linearity may in some instances be defined differently in NCCLS and GLP-compliant references. NCCLS document EP6-P 1998: 6(18) defines linearity as the measure of the degree to which a curve approximates a straight line. In contrast for GLP-compliant immunoassays, linearity has been defined using the International Conference on Harmonization (ICH) definition, "the condition in which test results are directly proportional to the concentration of analyte in the test sample."

[d] Accuracy (ICH definition) expresses the closeness of agreement between the value that is accepted either as a conventional true value or an accepted reference value and the value found (measured). Accuracy is sometimes termed "trueness." NCCLS document: EP 15-P, User demonstration of performance for precision and accuracy, Proposed Guideline 18(22), 1998.

Table 2 Common Analytical Issues for Validation of an In Vitro Quantitative
Immunoassay for a Novel Biomarker

Often analytical reagents are unique, not available widely, subject to lot-to-lot variability,
    and have limited stability.
Reference standard may be lacking or poorly characterized.
Reference standard may not be representative of endogenous biomarker.
Endogenous biomarkers often have innate molecular heterogeneity, including:
    Variable glycoslyation
    Size and charge variants
    Variants due to alternative splicing
    Precursors
    Homologous subunits
    Variant forms due to the disease state
    Analyte oliogomers
Endogenous nature of biomarkers complicates preparation of matrix-based calibrators and
    QC samples, and assessment of assay specificity.
Biomarker assays are prone to interference from matrix-based components, including
    binding proteins and proteases.
Disease state can introduce additional sources of assay "nonspecificity."

spectrometry are often available for small-molecular-weight biomarkers [10,11].
However, this is usually not the case for protein biomarkers, because of their
innate in vivo heterogeneity, low concentration in biological fluids, and the lack
of suitable comparison reference methods (Table 2) [9,12–15]. Recently isotope
dilution mass spectrometry has been used to develop reference methods for
hemoglobin A_{Ic} and apolipoprotein A-I [16,17]. Biosynthetic proteins, which are
used widely as reference standards in immunoassays, are known to often have
glycosylation patterns that are distinct from their endogenous counterparts.
Because of their heterogeneous nature, it is nearly impossible to prepare
glycoprotein standards that are identical to the endogenous circulating form
[9,10,18]. Ideally, the reference material should be "analytically" equivalent to
the endogenous biomarker. Therefore, for assays of protein biomarkers, we
recommend that experiments be conducted prior to prestudy validation to
appropriately characterize the purity and suitability of a reference material for use
as a calibration standard. For an assay standard obtained from an external source
or vendor, it is often necessary to obtain additional confirmatory data to
supplement the vendor's information. In some cases this can be problematic,
since labs that perform biomarker assays are frequently not well equipped to
adequately characterize reference standards. The lack of availability of a suitable
reference standard complicates both the design and conduct of experiments to

assess accuracy by "spike recovery." Consequently, assays that lack a suitable reference standard are regarded as providing a relative, but not a "truly" definitive, quantitative measure of the biomarker [8,9].

Another important analytical issue concerns the impact of the disease state on both the design and performance of a biomarker assay. The disease state can theoretically influence a biomarker assay in a number of ways. Usually, the biomarker of interest is present in biological matrices at some baseline level, and the presence of disease induces either an increase or decrease in its concentration or activity. The intersubject variability in biomarker concentrations is generally greater in patients with overt disease [8]. Since the biomarker concentration is a function of the disease state, it effectively dictates the concentration range over which the biomarker assay must be validated [8,9]. Therefore, it is often beneficial to conduct experiments to screen healthy subjects and individuals with disease prior to initiating prestudy validation to obtain a preliminary evaluation of the range of biomarker concentrations. For validation, we recommend that experiments are performed to investigate whether high concentrations of the biomarker, such as those produced by the disease, can cause a "high dose" hook (prozone effect) in the assay [19]. Noncompetitive immunoassays, such as sandwich ELISAs and immunoradiometric assays, are particularly prone to these effects. If a "high dose" hook effect is present, it is advisable to analyze test samples at multiple dilutions to ensure accurate reporting for samples having a high concentration of biomarker.

The disease state can influence the performance of a biomarker assay by introducing either "specific nonspecificity" or "nonspecific nonspecificity" [9,20]. A disease state may contribute to "specific nonspecificity" by promoting formation of variant forms of a biomarker that display variable cross-reactivity and nonparallelism relative to the reference standard. Since quantitative in vitro assays are often conducted without sample extraction, they are prone to interference from "nonspecific nonspecificity," which is also termed "matrix effects." In general, matrix effects produce a "false positive" in competitive assays and a "false negative" in noncompetitive assays. A disease state may contribute to nonspecific nonspecificity by influencing matrix components that are structurally unrelated to the analyte of interest, but that affect quantification of the biomarker [9,20]. It is noteworthy that it is often difficult to distinguish between a low concentration of the "endogenous" biomarker in a test sample and analytical bias introduced by the presence of matrix effects. Other possible sources for introduction of nonspecific nonspecificity include degradation, sequestration, or synthesis of the biomarker due to components in the test sample matrix.

The evaluation of analytical accuracy is often more problematic for assays of novel biomarkers than of xenobiotic drugs and routine clinical biomarkers (Table 1). For validation of bioanalytical assays for xenobiotic drugs, accuracy is usually defined in terms of "mean bias," which is the systematic difference (%R.E. or %recovery) between a mean measured result and the theoretical (nominal) true target value [9]. This typically involves measuring the drug of interest after "spiking" it into a biological matrix, and then computing the mean bias of the measured concentration relative to the target true (nominal) value [6–9]. In contrast, two somewhat different approaches are recommended for investigating the accuracy of assays for routine clinical biomarkers (Table 1). The first approach involves analyzing test samples by the newly developed analytical method and another established "gold standard" method. The results are then compared to determine whether a significant bias exists between the two methods. The second approach involves analyzing an official reference material by the newly developed method and then assessing recovery (bias). Some sources of reference samples with assigned values for routine clinical biomarkers include the National Institute of Standards and Technology (NIST), the College of American Pathologists (CAP), the National Reference System for the Clinical Laboratory (NRSCL), the Centers for Disease Control and Prevention (CDC), the National Institute for Biological Standards and Controls (NIBSC) in the United Kingdom, various proficiency/quality-control testing programs, and method manufacturers. The various approaches used to assess the accuracy of assays for routine clinical biomarkers are not practical for novel biomarkers. First, "gold standard" or comparator assays seldom exist for novel biomarkers, especially for those that are macromolecular. Thus, assessing accuracy by the method comparison approach is usually not an option. Second, because of their esoteric nature, "official" primary and secondary standards seldom exist for novel biomarkers. In the absence of an absolute reference material, spike recovery experiments provide only a "relative" measure of accuracy. Thus in the absence of absolute reference material, spike recovery experiments represent the most practical approach for evaluating the accuracy of a novel biomarker assay.

IV. STRATEGY FOR VALIDATION OF NOVEL BIOMARKER ASSAYS

The goal of biomarker analytical methods development and validation should be to "develop a valid (acceptable) method," rather than to simply "validate (accept) a developed method" [9,21]. For this reason, we advocate that assay development and validation to be viewed as a continuum of stages (Fig. 3). When feasible, it is preferable for the analyst to specify the minimum analytical performance specifications of a biomarker assay prior to initiating prestudy validation. Ideally, knowledge about the biological variability and mean difference in healthy and disease populations can be used to estimate the levels of analytical accuracy and precision needed to detect a change in biomarker concentration, due to either

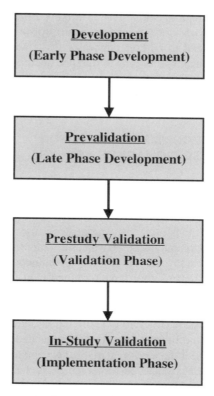

Figure 3 Stages of assay development and validation.

disease progression or drug treatment. However, owing to limited clinical experience in early-phase drug development, information is often lacking concerning the in vivo concentrations of a novel biomarker. This issue frequently complicates a priori definition of the concentration range required for prestudy validation. Obviously, as experience is gained in measuring a novel biomarker across multiple clinical studies, it may be beneficial to reevaluate the assay's requirements.

One strategy for addressing this dilemma is to conduct a pilot clinical study of the biomarker prior to initiating clinical studies to assess intrasubject and intersubject variability and the magnitude of the difference in biomarker concentrations between normal subjects and patients with disease. For prestudy validation of biomarker assays, we recommend caution in routinely applying the GLP-compliant limits of 15% and 20% (at the limits of quantitation) for accuracy and precision as the default values for method acceptance without seeking input from the clinician or pharmacokineticist. Unlike GLP-compliant bioanalytical assays for drugs, biomarker quantification requires different levels of analytical performance, which need to be either more or less rigorous depending on the specific biomarker. Thus, expectations for method accuracy and precision should be established after considering the different sources of variability and the mean difference in population levels of the biomarker.

V. VALIDATION RECOMMENDATIONS FOR CLINICAL-PHASE BIOMARKER ASSAYS

As described earlier, assays for novel biomarkers are often characterized by unique analytical challenges that make application of a strict set of validation procedures impractical. For these reasons, we advocate that novel biomarker assays be viewed as "GLP-similar" methods and that procedures for bioanalytical methods validation in guidance documents be viewed as a "framework" for designing and performing biomarker assay validation. We recommend that novel biomarker assays be validated according to a predetermined plan that defines the assay's target acceptance limits and the extent of prestudy validation. Each validation plan should be developed on a case-by-case basis that takes into account the analyte's properties, the type of assay, any analytical limitations, and the intended use of the biomarker data. For method acceptance, it is important to determine whether or not the validation results meet the a priori statistical acceptance criteria [9].

Table 3 lists the analytical performance characteristics that we consider important for prestudy validation of either a definitive or relative quantitative in vitro ligand-binding assay, including immunoassays. Figure 4A–C are schematics for the proposed activities in the prevalidation, prestudy validation and in-study validation phases of the validation of a definitive quantitative assay (immunoassay). These flowcharts assume that a high-quality reference standard is available for assay calibration. For quasi-quantitative and classification assays in which a suitable reference standard is lacking, fewer analytical performance characteristics are required for validation (Fig. 2, bottom boxes).

A. Development Phase

The development phase includes all activities that pertain to establishment of a "prototypic" quantitative assay [9,21]. For immunoassays, some of these activities include establishment of a suitable reference standard, evaluation of key reagents, characterization and selection of reagent antibodies, and defining the binding assay format. Details and recommendations for development of immunoassay methods have been reviewed extensively in previous publications [9,21–24].

Table 3 Prestudy Analytical Performance Parameters
for Quantitative In Vitro Immunoassays

Accuracy (mean bias)/total error
Imprecision:
    Intra-assay (within assay, or repeatability)
    Interassay (between assays or intermediate precision)
Sensitivity (LLOQ)
Specificity
Range (LLOQ/ULOQ)
Dilutional linearity
Parallelism
Standard stability
Matrix stability
    Freeze/thaw cycles
    Short-term stability
    Long-term stability

(a)

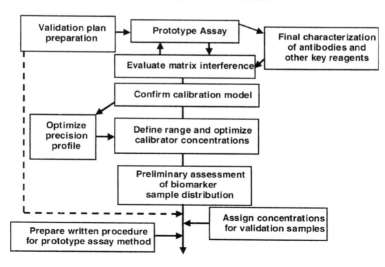

Figure 4 (a) Schematic of prevalidation phase for a quantitative immunoassay. (b)
Schematic of the prestudy validation phase for a quantitative immunoassay. (c) Schematic
of the in-study validation phase for a quantitative immunoassay.

(b)

Prestudy Validation Phase

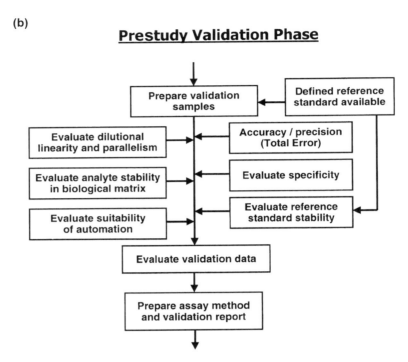

Figure 4   continued.

As noted previously, immunoassays and other ligand-binding assays are characterized by nonlinear calibration curves, in which the mean response is a nonlinear function of the analyte concentration, and the variance in replicate response measurements is a nonconstant function of the mean response [7,9,21,25]. The 4/5-parameter logistic model is the commonly acknowledged "reference" model for fitting immunoassay data [7,9,21,25]. We recommend that a suitable calibration model be determined during method development prior to initiating formal validation data. Inappropriate fitting of calibration response data will lead to undesired bias in the analytical method. Procedures have been described recently for optimizing and evaluating the acceptability of a nonlinear calibration model [9]. Recommendations included in the 2001 Food and Drug Administration (FDA) industry guidance for validation of immunoassays are consistent with these published calibration model recommendations. For example, the new FDA bioanalytical methods validation guidance recommends a minimum of six nonzero calibrators and use of the 4/5-parameter logistic

(c)

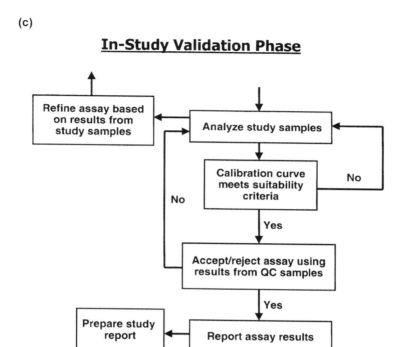

Figure 4   continued.

model, and acknowledges that calibrators outside the validated range (anchor points) can be beneficial for curve fitting [7].

One aspect of calibration that warrants careful consideration is the calibration curve matrix. Most guidance documents recommend that calibration curves be prepared in the same species matrix as the test samples [6,7]. This helps to ensure consistency in the concentration–response relationship between the calibrators and the analyte in test samples. However, unlike xenobiotic drugs, the endogenous nature of biomarkers greatly complicates the use of unaltered sample matrix for preparation of calibrators. A variety of strategies can be used to minimize or eliminate interference from endogenous biomarkers [9]. If a physiological buffer is used as a surrogate matrix for preparing calibrators, it is important to investigate whether the buffer introduces a bias in the assay. For immunoassays the 2001 FDA guidance for bioanalytical methods validation provides the flexibility of preparing calibrators in "an alternate matrix of equivalent performance" [7].

B. Prevalidation Phase

In general, the prevalidation phase (i.e., method optimization phase) for a quantitative novel biomarker assay resembles that of any immunoassay [9]. However, as mentioned previously, the diverse and case-by-case nature of assays for novel biomarkers makes application of a strict set of validation procedures problematic. Therefore, upon establishment of a prototypic assay method, we recommend the development of a formalized validation plan that defines the scope and purpose of the assay, provides appropriate background information, and highlights any practical or scientific issues that may affect the design of prestudy validation. Additionally, the plan should include details about the validation experiments to be performed and, when feasible, include a priori criteria for method acceptance. Depending on the extent of prior knowledge about the biomarker, it may be appropriate to develop the validation plan either early in the prevalidaiton phase or immediately prior to initiating prestudy validation experiments (Fig. 4B).

During the prevalidation phase it is useful to perform a preliminary assessment of the biomarker's distribution in individuals with and without disease to establish the concentration range over which the assay needs to be validated. If the range of calibrators is inadequate for bracketing the expected range of biomarker concentrations, dilutional linearity experiments should be planned for prestudy validation to demonstrate validity of sample dilution as a means for bringing the biomarker's concentrations "in range."

One important step during the prevalidation phase involves defining the biomarker concentrations for validation samples that will be used to assess accuracy and precision. Recently, Findlay et al. described a simplified procedure for obtaining preliminary estimates of the lower and upper limits of quantification [9]. For convenience, we recommend preparing validation samples at the following five concentration levels, one at the anticipated lower limit of quantification (LLOQ), one about two to four times the LLOQ, one near the assay's midrange (on logarithmic scale), one at about 70–80% of the anticipated upper limit of quantification (ULOQ), and one at the anticipated ULOQ. The endogenous nature of biomarkers can present an issue in the preparation of both prestudy validation samples and in-study quality control (QC) samples. Thus, it is often necessary to devise a matrix-specific strategy to permit assessment of analytical accuracy and precision at the LLOQ prior to initiating prestudy validation experiments [9]. The final step in the prevalidation phase is preparation of a preliminary written method that describes in detail the assay that will be evaluated during prestudy validation. For obvious reasons, it is not appropriate to make methodological changes upon initiation of prestudy validation experiments.

C. Prestudy Validation Phase

We recommend the analytical performance characteristics described in Fig. 4B be investigated during prestudy validation. Our default recommendations for target acceptance limits and statistical acceptance criteria of in vitro quantitative assays to quantify novel biomarkers are reported in the recent publication by Findlay et al. [9] As stated previously, these default limits and criteria should be modified on a case-by-case basis after consideration of the different sources of variability, mean difference in population levels of the biomarker, any analytical limitations, and the intended purpose of the assay. Additionally, we support including an assessment of "total analytical error," as part of the prestudy statistical acceptance criteria for evaluating accuracy and precision [7,9]. These default recommendations are more liberal than the method acceptance criteria outlined in FDA guidances for bioanalytical methods validation [6,7]. We believe these criteria are justified, because novel biomarkers are often macromolecular and known to have unique analytical issues (Table 2). The publication by Findlay et al. and the new FDA guidance agree that at least six independent assay runs should be made during prestudy validation to assess accuracy (mean bias) and intermediate precision. It is important to emphasize that during prestudy validation it is inappropriate to reject assay runs, except in rare instances where technical or other inexplicable factors are obvious.

As outlined in Table 3, biomarker stability in the biological matrix should be determined during prestudy validation. However, the same analytical limitations that impact the development and validation of biomarker assays also affect the conduct of experiments to assess the matrix stability of a biomarker. Thus, the design of stability experiments usually involves following a change in either an endogenous or a "spike" concentration of the novel biomarker. When appropriate, we recommend that biomarker stability be evaluated in pools of matrix from both normal subjects and patients with disease. If a suitable biomarker standard is available, this permits assessing the change in biomarker concentrations in "spiked" matrix samples. The number of freeze–thaw cycles should reflect the expected handling of the samples and should be minimized, as macromolecule biomarkers are usually less stable than conventional drugs. We recommend that stability assessments be performed in "unaltered" test sample matrix, as any effort to remove the endogenous analyte will also likely introduce changes to the biological matrix [9]. We recommend that a priori criteria be used to define the time interval over which the biomarker is determined to be stable. The recent FDA guidance makes specific recommendations for the design of experiments to assess stability in the matrix [7].

Although it is beyond the scope of this chapter, another important aspect of novel biomarker stability assessment concerns optimizing the "preanalytical" conditions used for collecting, processing, and storing test samples. The matrix concentrations of some biomarkers, such as cytokines, are known to be highly dependent on sample-handling conditions, including the type of anticoagulant, centrifugation conditions, and storage temperature. Also, information regarding a biomarker's metabolic pathways may be useful to ascertain whether biotransformation may lead to a change in concentration. Quality control charts can be used to prospectively follow both sample integrity and analyte stability [26,27].

D. In-Study Validation Phase

Criteria for accepting or rejecting assay runs of test samples depend on the measurement category of the biomarker assay (Fig. 4). Decisions about the acceptability of an assay run are based largely on the results for quality control samples (QCs). For in vitro quantitative assays that employ a reference standard (i.e., definitive and relative), we recommend the use of either a "4–6- × " rule [6,7,9] or control charts [26,27]. Owing to analytical issues inherent in assays of novel biomarkers, we support recommendations in recent publications to make the "4-6-20" rule for bioanalytical assays of drugs more lenient for biomarker quantification [9,28]. When feasible, it is convenient to use three levels of the validation samples as QCs to monitor in-study assay runs (i.e., one within two to four times the LLOQ, one near the assay's midrange, and one near 70–80% the ULOQ) [7]. If the biomarker concentration is anticipated to exceed the assay's ULOQ in a majority of test samples, it may be appropriate to include one or more additional quality control samples in this concentration range [9]. For in vitro quasi-quantitative assays that do not employ a reference standard, control charts, which are used to monitor the analytical response of QC samples, are the preferred method for accepting or rejecting assay runs [26,27]. Control charts are constructed using the mean response and imprecision for pools of quality control samples that span the anticipated range of biomarker levels in clinical samples. In general, run acceptance or rejection for a classification assay is based on results from positive and negative controls.

E. Commercial Immunoassay Kits

Commercial assay kits warrant further discussion, as they are used often during clinical studies for quantification of novel biomarkers. Commercial assay kits can minimize the time for method development, are usually cost-effective, and can facilitate sample throughput. These features make commercial kits attractive for use in quantifying biomarkers during clinical investigation of new drug entities. However, it is important to recognize that the intended use for commercial kits is typically diagnostic, usually with a cutoff point for diagnosis of disease. Therefore, in some situations issues can exist in regard to sensitivity (LLOQ), range, and specificity when commercial kits are used to measure a novel biomarker in support of clinical drug study. For example, the kit's accuracy and precision may not be adequate to measure the range of biomarker concentrations found in test samples. It is not uncommon for a clinical drug study to require greater sensitivity, because often the goal is to quantify baseline levels of the biomarker. Additionally, in some cases the drug of interest or other coadministered medications can cause analytical interference. Also, the commercial kit may not perform well with samples from patients with overt disease. For these reasons, when a commercial kit is used to provide key pharmacodynamic assessments for either safety or efficacy, we recommend that the kit be validated fully prior to use. Particularly for biomarker quantification, cross-reactivity (spiking) or other experiments should be conducted to document a lack of interference from the investigational drug and its putative metabolites. For any commercial kit that is intended for research purposes only (not FDA-approved), we recommend that a pure standard (if available) be used to compare calibration standards across different lots and between kits from different manufacturers. As is the case with any biomarker assay, the design and conduct of the kit validation will depend upon the technical feasibility and other limitations.

VI. CASE STUDIES TO ILLUSTRATE ANALYTICAL VALIDATION OF NOVEL BIOMARKER ASSAYS

In this section we describe three examples of assays for in vitro quantification of novel biomarkers that have been used as biological endpoints during drug development. From an assay categorization perspective (Fig. 2), these include a definitive quantitative assay for small-molecular-weight analytes and two relative quantitative assays for macromolecular analytes.

A. Definitive Quantitative Assay of Small-Molecule Analytes

1. Xanthine, Hypoxanthine, and Uric Acid

Xanthine oxidase converts hypoxanthine (HX) to xanthine (X) and then to uric acid (UA). High circulating concentrations of UA (hyperuricemia) result in deposition of the urate crystal, which can lead to joint inflammation (gout) and renal impairment in the kidney. A high-performance liquid chromatographic (HPLC) method was developed to quantify X, HX, and UA in serum and urine as biomarkers to evaluate disease progression and drug pharmacodynamics. Because of the presence of a large number of interfering polar components in urine, a LC–MS–MS assay was developed and validated to provide better analytical selectivity [29]. The concentration of endogenous analytes was in the µM range and varied among individuals. Standards were prepared by "spiking" a buffer with commercially obtained reference compounds. Low, middle, and high QCs (i.e., validation samples) were prepared by spiking known concentrations into serum. For urine, the three spiking QC levels were prepared in a buffer solution, and a fourth QC was made from pooled urine containing the endogenous compounds. Internal standards (I.S.) were added directly to the samples. Each urine test sample was diluted ninefold with 47 mM potassium phosphate (monobasic) and injected directly onto either the HPLC or LC–MS–MS. To remove serum proteins prior to injection onto an HPLC system, each serum sample was diluted four times and passed through an ultrafiltration membrane with a ~30 kd-molecular-weight cutoff. The chromatographic conditions included a C_{18} reversed-phase column with a mobile phase of 47 mM potassium (monobasic) phosphate and detection by ultraviolet. The internal standard (I.S.), uridine, had a run time of 13 and 21 min for serum and urine, respectively. A column-switching procedure, which incorporated a secondary mobile phase of 50% acetonitrile in 47 mM potassium (monobasic) phosphate, was used to eliminate interference from urine late-eluting peaks. For quantification by LC–MS–MS, ^{15}N-labeled analogs of each analyte were synthesized for use as internal standards. An AB/MDS Sciex API 3000 was used with Turbo IonSpray interfaced to two diisopropyl-C_{14} amide analytical columns with a dual-column switch setting. Negative ions were monitored in the multiple-reaction mode. The run time for LC–MS–MS was 5.5 min.

   Literature information was used to establish the range of potential analyte concentrations in clinical samples. The levels of X and HX were typically much lower than that of UA. Serum concentrations were lower than those in urine. Samples from 17–20 persons were collected and screened to help confirm the appropriate standard curve ranges. The standard and QC sample concentrations in Table 4 were selected to cover the expected concentration range in study samples. Since urine UA levels in screened samples were found to have high concentrations, the LLOQ for the standard range of the subsequently developed LC–MS–MS method was adjusted to be 10 times higher than for the HPLC method. Serum QC samples were prepared by spiking known amounts of each analyte into a pool of serum from persons who had the lowest concentrations of endogenous analytes. Because of the likely contribution of basal analyte at a concentration below the lowest standard, the low-QC concentration was assigned the mean concentration determined during prestudy validation. The middle- and high-QC samples were assigned the nominal spiked concentration values. In contrast to serum, urine samples had relatively high concentrations of endogenous analyte. This precluded use of standard addition at multiple levels as a means for preparing QC samples. Therefore, a single QC sample was prepared from a pool of human urine and the mean concentration was measured

Table 4 Case Study: Design of Xanthine, Hypoxanthine, and Uric Acid Standards and QC Samples

	Serum (μmol/L)		Urine (μmol/L)			
					LC–MS–MS	
	HPLC standard range	HPLC QC concentrations[a]	HPLC standard range	HPLC QC concentrations[b]	standard range	LC–MS–MS QC concentrations[b]
X	0.2–20	0.57, 3, 15	10–1000	30, 49,[c] 200, 750	10–1000	30, 82.3,[c] 200, 750
HX	0.2–20	0.57, 3, 15	10–1000	30, 194,[c] 200, 750	10–1000	30, 62.9,[c] 200, 750
UA	10–1000	29.6, 150, 750	10–4500	23.6, 500, 2650,[c] 3375	100–4500	300, 1250, 1611,[c] 3375

[a] QC samples were prepared by spiking known amounts of analytes into a pool of sera that had the lowest observed analyte concentrations. The low QC values were determined by the validation mean, while the mid- and high QCs were nominal values.
[b] Nominal values in buffer solution.
[c] Validation mean concentration for this QC, which was prepared from pooled urine.

during prestudy validation. Three additional levels of QC samples were prepared in buffer.

Five to eight prestudy validation runs with six determinations per batch were conducted to assess the analytical performance. For each method, the accuracy (mean bias) and of interassay coefficient of variation (CV) for validation samples were well within the a priori 15% acceptance limits. Overall, the maximum observed bias was 8.3% and the maximum CV was 9.2%. Accuracy could not be assessed for the low-serum-and-urine QC samples, because these were assigned the validation mean as their target concentration. For in-study assay runs, acceptance was assessed by examining the results for QC samples. The run acceptance criteria (4-6-15 rule) was the same as that used to support bioanalytical assays of the investigational drug (i.e., 67% of QC results had to be within 15% of the target values).

Since some urine specimens had very high urate concentrations, the uric acid formed insoluble crystals upon storage at temperature lower than the body temperature. Alkaline was added to assure solubility over the 24-h urine collections. Because of the similarity of xanthine and hypoxanthine to caffeine in the diet, the method had to be shown to be selective against caffeine. During the method change for urine, clinical and spiked urine sample results from both methods were compared and found to be within 15% of one another. The LC–MS–MS method provided better selectivity and signal to noise and a faster throughput than the HPLC method.

B. Relative Quantitative Assay of a Macromolecular Analyte

1. Vascular Endothelial Growth Factor

Vascular endothelial growth factor (VEGF) is an example of an endogenous peptide that is important as a biomarker. VEGF is an endothelial cell mitogen and angiogenic factor that plays an important role in tumor vascularization and growth. Additionally, the intensity of the angiogenic response, as measured by the degree of tumor vascular density, correlates with poor prognosis in several types of cancers [30,31]. Toi et al [32,33]. used immunocytochemical means to show a close association between VEGF expression and the increase in tumor microvasculature density in breast cancer and the importance of reduced circulating VEGF levels as a predictor of stable disease. In preclinical models with human tumor xenografts, plasma VEGF is often elevated and directly correlated with tumor mass. Consequently, VEGF is often a biomarker that is monitored in clinical trials for cancer chemotherapy agents. VEGF is expressed as four different peptides containing 121, 165, 189, and 206 amino acid residues. $VEGF_{121}$ and $VEGF_{165}$ are secreted and are found in the circulation, whereas $VEGF_{189}$ and $VEGF_{206}$ are mostly associated with heparin in the extrecelluar matrix [34]. We will describe an ELISA method developed for quantifying plasma VEGF to support clinical trials of cancer chemotherapy agents. Since immunological reagents were available commercially, our principal task was to develop and validate an assay based on reagents that were purchased in bulk.

The assay that was developed was based on a commercially available immunoassay from R&D Systems, Minneapolis, Minnesota. The kit assay employs monoclonal antisera to $VEGF_{165}$ coated on the plate, polyclonal $VEGF_{165}$ antisera conjugated to horseradish peroxidase (HRP) as the second reporter antibody, and $VEGF_{165}$ as standard. To decrease background signals and material cost, we purchased the primary monoclonal antibody and standard in bulk from R&D Systems and used polyclonal anti-$VEGF_{165}$ antisera from Endogen (Woburn, MA) to complete the ELISA "sandwich." The final component in the assay format is a goat antirabbit IgG polyclonal conjugated to HRP. As $VEGF_{121}$ is present in human blood, we established this form was 100% cross-reactive with antisera against $VEGF_{165}$. Consequently, we report the measured values as $VEGF_{165}$ equivalent concentration and made sure that our customers, the biologists, pharmacokineticists, and clinical scientists, were aware of this fact. Since biomarkers often exist in multiple forms, it was important to define the reference standard being used in each assay. In this case, it was defined as a peptide of $VEGF_{165}$. However, in other situations, it could be a mixture of peptides of known proportions (e.g., the PSA tumor marker [35]). Our recommendation was to report results in terms of molar concentrations (e.g., nM), rather than reporting "equivalents" based on either mass or volume. The conversion factor should be available in the report for the data users' convenience.

The kit's standard curve ranged from 39 to 1000 pg/mL in heparinized plasma. Using the Endogen second antibody, the range of the curve was 78–10,000 pg/mL. To prepare standards and QCs, 15 separate lots of heparinized plasma from normal adults were analyzed with the kit assay for $VEGF_{165}$ equivalent concentration. Thirteen of 15 lots were found to have undetectable levels (≤ 39 pg/mL) and were pooled for use as control matrix. The calibration standard was prepared for each assay in control matrix. During prestudy validation, standards ranging from 19.5 to 40,000 pg/mL were used to define the acceptable range. A four-parameter logistic model was used to fit calibration curves. Standards at 19.5 and 39 and 20,000 and 40,000 pg/mL were eliminated from the curve fit, because the relative standard deviation (RSD) and/or relative error (RE) usually exceeded 25%. QCs were prepared at 100 and 500 pg/mL and 45 ng/mL (diluted 10-fold before analysis as a dilution QC) and determined in six assays (two each for 3 days). Interassay precision based on these determinations was 12.7%, 8.0%, and 5.8% RSD for the low, medium, and high QCs, respectively. Assay accuracy for the same QCs was -4.0%, -11.0%, and 0.71% RE, respectively. Analyte stability was determined to be acceptable for 2 years at

$- 80°C$ and through three cycles of freezing–thawing, as results for QC samples were within 20% of their nominal values. Assay reagents were shown to be stable for at least 2 weeks when stored refrigerated, at $2-8°C$.

Even though analysis of the control matrix lots used in preparing the pooled control showed the endogenous $VEGF_{165}$ equivalent concentrations to be lower that 39 pg/mL, undoubtedly some basal concentration of analyte was present. The basal level of analyte did not affect accuracy of QC sample measurements, because standards were prepared in the same matrix pool. However, the concentration of an unknown test sample may not be accurate. In other words, although this design may yield a highly precise assay, the method is most likely characterized by a small bias due to the presence of endogenous concentrations of analyte in the calibrator matrix. Despite this, the ELISA is still a very important tool in measuring changes in VEGF response during disease progression and pharmaceutical interventions. In summary, owing to issues pertaining to assay calibration, this ELISA would be categorized as a relative quantitative assay.

C. Relative Quantitative Assay of a Receptor Activity

1. Inhibition of Pan ErbB Receptor Tyrosine Kinase

When growth factors bind to receptors, they initiate a series of events that culminate in some intracellular response [36]. In tumor cells, these responses can be aberrant proliferative responses resulting in tumor growth and disease progression [37,38]. When drug therapy is developed that targets a specific receptor, a prime marker of mechanism of action would be to measure the drug's effect on the receptor. This case study will describe how to develop and characterize an analytical method that measures the response of a receptor to the effect of a drug.

The ErbB family of receptors include ErbB-1 (epidermal growth factor receptor, EGFr), ErbB-2 (HER2/neu), ErbB-3 (HER3), and ErbB-4 (HER4) [39,40]. The receptors have an extracellular ligand-binding domain, a transmembrane domain, and, except for ErbB-3, an active intracellular tyrosine kinase [41,42]. The ErbB receptors exist as monomers until activated by ligand binding, which results in dimerization with one of the four possible receptors. This dimerization results in activation by ATP-dependent receptor phosphorylation at one of several tyrosine residues near the C-terminus. The phosphorylation leads to a signal transduction cascade that results in uncontrolled tumor growth [43]. CI-1033 is an investigative compound that inhibits the tyrosine kinase activity in ErbB-1, -2, and -4. To measure the effect of CI-1033, a commercially produced ELISA assay to quantitate total ErbB-1 or EGRr has been adapted to measure both total EGFr and the amount of phosphorylated EGFr.

The assay developed was based on an ELISA for total EGFr (EGF-R ELISA, Oncogene Research Products, Boston, MA). Monoclonal antibodies to two different extracellular sites of EGFr are used for the primary, 96-well coated antibody and the secondary reporter antibody. The reporter antibody is conjugated to biotin and completion of the ELISA "sandwich" is accomplished by adding avidin-conjugated horseradish peroxidase (HRP). The binding reaction is detected by HRP-catalyzed conversion of tetramethylbenzidine (TMB) to its colored product. To quantitate the phosphorylated receptor, biotin-conjugated monoclonal antibody (4G10) from Upstate Biotechnology, Lake Placid, NY, that recognizes the phosphorylated tyrosine residues is added as the reporter antibody. The final response is the same conversion of TMB to colored product as in the total assay. The standard curve for the total receptor is 3.1–100 femtomoles (fmol)/mL and the phosphorylated assay is 71–2770 fmol/mL.

The ELISA assay (Oncogene) for the total receptor was used in accordance with the information included in the kit insert. EGFr is 170 kilodaltons (kd) in mass but may also exist as the 110-kd truncated extracellular portion only. Because the primary and reporter antibodies are both directed against the extracellular domain, the cross-reactivity of the assay for the 110-kd fragment is 100% relative to the complete 170-kd peptide. The kit was designed to be used for tissue or tumor lysate, cell culture supernatants, or plasma, and since either forms of the peptide may be present, the results are reported as fm/mL. Lyophilized standard material provided in the kit is the complete 170-kd peptide, prepared from human A431 tumor epidermoid carcinoma cell lysates and calibrated against independent immunoaffinity purified EGFr. The curve consists of 100, 75, 50, 25, 12.5, 6.3, and 3.1 fmol/mL standards and data are fitted using a four-parameter logistic model. Since pure, characterized total or phosphorylated receptor was not available, QC samples, as defined in the classic sense, could not be prepared. However, to ensure that the assay was reproducible, quality assurance samples (QAs) were prepared. One QA consisted of an in-house A431 xenograft tumor lysate and a second QA was prepared from commercially available EGFr (Sigma, St. Louis, MO) diluted in tumor lysate buffer. Aliquots of both QA samples were prepared and stored frozen at $-80°C$. For each analysis QA samples were thawed for each analysis and diluted appropriately. Values of the two QAs, determined during assay characterization and routine analysis, are as follows: in-house sample, mean $= 161,000$ fmol/mL, 18.4% CV, $N = 25$ assays; Sigma sample, 1,050,000 fmol/mL, 10.5% CV, $N = 8$ assays.

The main analytical challenge for developing the ELISA for the activated phosphorylated receptor is the inability to obtain certifiably pure material. In addition, there is the added complication that there are four or five possible phosphorylated sites on each receptor with no currently available method to determine the degree of phosphorylation at each potentially phosphorylated tyrosine residue. Consequently, the standard curve for the ELISA is as follows: (1)

the standard added is the same lyophilized human A431 epidermoid carcinoma xenograft lysate that is used in the total receptor assay; (2) the second, reporter antibody is specific for phosphorylated tyrosines residues [Upstate Biotechnology Inc. (UBI), Lake Placid, NY]; and (3) the amount of standard is in effect an equivalent concentration of total receptor. Since the absolute amount of the phosphorylated form cannot be determined, results from the assay are reported as the ratio of active to total.

As in the case of the ELISA for total receptor, to ensure that the assay was performing reproducibly from assay to assay, QAs were prepared. One QA consisted of the in-house A431 tumor lysate described earlier while the second QA was prepared with material from UBI. The QA samples were aliquoted, stored frozen at $-80°C$, thawed for each analysis, diluted to an appropriate concentration, and analyzed. Values of the two QAs, determined during assay characterization and routine analysis, are as follows: in-house sample, mean $=5670$ fmol/mL, 23.4% CV, $N=26$; UBI sample, mean $=28,900$ fmol/mL, 21.7% CV, $N=11$. Lysate sample storage stability was determined to be acceptable for 6 months at $-80°C$ as results for QA samples remained within 20% of initially determined values.

These types of receptor "activity" assays are not easily validated in the classic GLP sense. However, they have provided a powerful tool for assessing the ability of EGFr inhibitor drugs to act on the expected target of action. In preclinical A431 human tumor xenograft models, CI-1033 has been shown to inhibit EGFr phosporylation and this inhibition is directly correlated to inhibition of tumor growth. Currently, the assays are being employed to determine degree of EGFr inhibition in early phase I clinical trials. In addition, a similar assay for total and phosphorylated ErbB-2 (HER2/*neu*) was used in a study to determine the effect of the overexpression of this receptor in breast cancer [44].

VII. CONCLUSIONS AND FUTURE CONSIDERATIONS

Analytical method development and validation for novel biomarkers depend ultimately on the intended use of the analytical measurements. Generally, the extent of analytical validation increases as a new drug entity progresses in the drug development continuum from discovery, to early-phase evaluation (e.g., preclinical and early clinical phases), clinical evaluation (phases II and III), and postmarketing studies (phase IV). Currently, in-depth guidance is lacking for validation of assays for novel biomarkers. Therefore, method validation should be specified in a formalized plan and conducted according to the type of analytical measurement (e.g., definitive and relative quantitative, or quasi-quantitative). Biomarker concentrations in samples from normal persons and patients with disease should be used to establish the range over which the assay needs to be validated. For validation, a priori method acceptance criteria should be established by the bioanalytical scientist in consultation with data recipients, such as clinicians and pharmacokineticists, to ensure the biomarker assay will be acceptable for its intended application. We have presented our recommendations on quantitative biomarker measurements and case studies to illustrate our proposed process for method development and validation.

Currently, immunoassays comprise the principle technology for quantification of protein and other macromolecular biomarkers. Chromatographic and LC–MS methods are often used to measure small-molecular-weight biomarkers, particularly ones that are nonpolar. In some situations ligand-binding assays may be useful for measuring small-molecular-weight biomarkers, if they are highly polar or amphoteric. Various types of hyphenated MS methods and matrix-assisted laser desorption ionization time of flight MS (MALDI-TOF MS), coupled with capillary electrophoresis and affinity chips, are becoming important new quantitative tools for discovery, characterization, and quantification of macromolecular biomarkers. Assays for genomic biomarkers (SNPs and mRNAs) are becoming important tools in drug development. Assay technologies that can link functional activity to immuno- (or mass-) quantitation on the same sample should be possible with chip technology using receptors, DNA or mRNA, and antibodies. Technological advances in automation, multiplexing multianalyte assays, such as Luminex [45,46], and microfluidic technologies will facilitate quantification of novel biomarkers. However, these new technologies also create new issues to consider during method development and validation. The future availability of new state-of-the-art technologies for quantification of novel biomarkers will require cooperation by analytical scientists, developers of diagnostic tests, clinicians, pharmacokineticists, and regulatory agencies to devise and implement a dynamic, yet standardized and systematic, approach for analytical method validation.

REFERENCES

1. Biomarkers and surrogate endpoints in clinical research: definitions and conceptual model in biomarkers and surrogate endpoints. In *Proceedings of the NIH–FDA Conference*, Apr 15–16, 1999; Downing, G.J., Ed.; Elsevier: New York, 2000; *3*.
2. Levy, G. Mechanism-based pharmacodynamics modeling. Clin. Pharmacol. Ther. **1994**, *56*, 356–358.
3. Peck, C.C.; Barr, W.H.; Benet, L.Z.; et al. Opportunities for integration of pharmacokinetics, pharmacodynamics, and toxicokinetics in rational drug development. Pharm. Sci. **1992**, *81*, 600–610.
4. Colburn, W.A. Selecting and validating biologic markers for drug development. J. Clin. Pharmacol. **1997**, *37*, 355–362.

5. Vesell, E.S. Advances in pharmacogenetics and pharmacogenomics. J. Clin. Pharmacol. **2000**, *40*, 930–938.
6. Shah, V.P.; Midha, K.K.; Dighe, S.; et al. Analytical method validation: bioavailability, bioequivalence, and pharmacokinetic studies. J. Pharm. Sci. **1992**, *81*, 309–312.
7. Guidance for industry on bioanalytical method validation: availability. Fed. Reg. **2001**, *66*, 28526–28527.
8. Lee, J.W.; Hulse, J.D.; Colburn, W.A. Surrogate biochemical markers: precise measurement for strategic drug and biologics development. J. Clin. Pharmacol. **1995**, *35*, 464–470.
9. Findlay, J.W.A.; Smith, W.C.; Lee, J.W.; et al. Validation of immunoassays for bioanalysis: a pharmaceutical industry perspective. J. Pharm. Biomed. Anal. **2000**, *21*, 1249–1273.
10. Stenman, U.-H. Immunoassay standardization: is it possible, who is responsible, who is capable? Clin. Chem. **2001**, *47*, 815–820.
11. Stockl, D.; Franzini, C.; Kratochvila, J.; et al. Analytical specifications of reference methods: compilation and critical discussion. Eur. J. Clin. Chem. Biochem. **1996**, *34*, 319–337.
12. Whicher, J.T. Calibration is the key to immunoassay but the ideal calibrator is unattainable. Scand. J. Clin. Lab. Investig. **1991**, *205* (Suppl.), 21–32.
13. Mire-Sluis, A.R.; Gaines-Das, R.; Thorpe, R. Immunoassays for detecting cytokines: what are they really measuring? J. Immunol. Methods **1995**, *186*, 157–160.
14. Zerfaoui, M.; Ronin, C. Glycosylation is the structural basis for changes in polymorphism and immunoreactivity of pituitary glycoprotein hormones. Eur. J. Clin. Chem. Clin. Biochem. **1996**, *34*, 749–753.
15. Tsang, M.L.-S.; Weatherbee, J.A. Cytokine assays and their limitations. Aliment. Pharmacol. Ther. **1996**, *10* (Suppl. 2), 55–61.
16. Barr, J.R.; Maggio, V.L.; Patterson, D.G., Jr.; et al. Isotope dilution-mass spectrometric quantification of specific proteins: model application with apolipoprotein A-I. Clin. Chem. **1996**, *42*, 1676–1682.
17. Kobold, U.; Jeppsson, J.-O.; Dulffer, T.; et al. Candidate reference methods for hemoglobin Alc based on peptide mapping. Clin. Chem. **1997**, *43*, 1944–1951.
18. Ekins, R. Immununoassay standardization. Scand. J. Clin. Lab. Investig. **1991**, *205* (Suppl.), 33–46.
19. Ryall, R.G.; Story, C.J.; Turner, D.R. Reappraisal of the causes of the hook effect in two-site immunoradiometric assays. Anal. Biochem. **1982**, *127*, 308–315.
20. Chard, T. An introduction to radioimmunoassay and related techniques. In *Laboratory Techniques in Biochemistry and Molecular Biology*, 3rd Ed.; Burdon, R.H., Van Knippenberg, P.H., Eds.; Elsevier: New York, 1987; Vol. 6, 175–193.
21. Smith, W.C.; Sittampalam, G.S. Conceptual and statistical issues in the validation of analytic dilution assays for pharmaceutical applications. J. Biopharm. Stat. **1998**, *8*, 509–532.
22. Findlay, J.W.A.; Das, I. Some validation considerations for immunoassays. J. Clin. Ligand Assay **1997**, *20*, 49–55.

23. Findlay, J.W.A.; Das, I. Validation of immunoassays for macromolecules from biotechnology. J. Clin. Ligand Assay **1998**, *21*, 249–253.

24. Khan, M.N. Immunoassay in drug development arena: an old player with new status. J. Clin. Ligand Assay **1999**, *22*, 242–245.

25. Dudley, R.A.; Edwards, P.; Ekins, R.P.; et al. Guidelines for immunoassay data processing. Clin. Chem. **1985**, *31*, 1264–1271.

26. Westgard, J.O.; Barry, P.L.; Hunt, M.R.; et al. A multi-rule shewart chart for quality control in clinical chemistry. Clin. Chem. **1981**, *27*, 493–501.

27. Westgard, J.O.; Klee, G.G. Quality assurance. In *Textbook of Clinical Chemistry*; Tietz, N.W., Ed.; WB Saunders: Philadelphia, 1986; 424–458.

28. Shah, V.P.; Midha, K.K.; Findlay, J.W.A.; et al. Bioanalytical method validation—a revisit with a decade of progress. Pharm. Res. **2000**, *17*, 1551–1557.

29. Cooper, N.; Brown, P.; Xu, L.; et al. Bioanalysis of xanthine, hypoxanthine and uric acid in urine by HPLC and LC/MS/MS methods: method comparison for pharmacodynamic biomarkers. AAPS Meeting Abstract, Denver, CO, 2000.

30. Schlaeppi, J.-M.; Eppenberger, U.; George, M.-B.; Küng, W. Chemiluminescence immunoassay for vascular endothelial growth factor (vascular permeability factor) in tumor–tissue homogenates. Clin. Chem. **1996**, *42*, 1777–1784.

31. Braybrooke, J.P.; O'Bryne, K.J.; Propper, D.J.; et al. A phase II study of razoxane, an antiangiogenic topoisomerase II inhibitor, in renal cell cancer with assessment of potential surrogate markers of angiogenesis. Clin. Cancer Res. **2000**, *6*, 4697–4704.

32. Toi, M.; Takayanagi, S.; Tominaga, T. Association of vascular endothelial growth factor expression with tumor angiogenesis and with early relapse in primary breast cancer. Jpn. J. Cancer Res. **1994**, *85*, 1045–1049.

33. Toi, M.; Inada, K.; Suzuki, H.; Tominaga, T. Tumor angiogenesis in breast cancer: its importance as a prognostic indicator and the association with vascular endothelial growth factor expression. Breast Cancer Res. Treat. **1995**, *36*, 193–204.

34. Rodriguez, C.R.; Fei, D.T.; Keyt, B.; Baly, D.L. A sensitive fluorometric enzyme-linked immunosorbent assay that measures vascular endothelial growth factor$_{165}$ in human plasma. J. Immunol. Methods **1998**, *219*, 45–55.

35. Rafferty, B.; Rigsby, P.; Rose, M.; et al. Reference reagents for prostate specific antigen (PSA): establishment of the first international standards for free PSA and PSA (90:10). Clin. Chem. **2000**, *46*, 1310–1317.

36. Huang, S.-M.; Harari, P.M. Epidermal growth factor receptor inhibition in cancer therapy. Investig. New Drugs **1999**, *17*, 259–269.

37. Woodburn, J.R. The epidermal growth factor and its inhibition in cancer therapy. Pharmacol. Ther. **1999**, *82*, 241–250.

38. Salamon, D.S.; Brandt, R.; Ciardiello, F.; et al. Epidermal growth factor-related peptides and their receptors in human malignancies. Crit. Rev. Oncol. Hematol. **1995**, *19*, 183–232.

39. Fernandes, A.M.; Hamburger, A.W.; Gerwin, B.I. Dominance of ErbB-1 heterodimers in lung epithelial cells overexpressing ErbB-2: both ErbB-1 and ErbB-2 contribute significantly to tumorogenicity. Am. J. Respir. Cell Mol. Biol. **1999**, *21*, 701–709.

40. Maghal, N.; Sternberg, P.W. Multiple positive and negative regulators of signalling by the EGF-receptor. Curr. Opin. Cell Biol. **1999**, *11*, 190–196.

41. Riese, D.J.; Stern, D.F. Specificity within the EGF family/ErbB receptor family signalling network. Bioassays **1998**, *20*, 41–48.

42. Hackel, P.O.; Zwick, E.; Prenzel, N.; et al. Epidermal growth factor receptors: critical mediators of multiple receptor pathways. Curr. Opin. Cell Biol. **1999**, *11*, 184–190.

43. Vincent, P.W.; Bridges, A.J.; Dykes, D.J.; et al. Anticancer efficacy of the irreversible EGFr tyrosine kinase inhibitor PD0169414 against human tumor xenografts. Cancer Chemother. Pharmacol. **2001**, *45*, 231–238.

44. Eppenberger-Castori, S.; Kueng, W.; Benz, C.; et al. Prognostic and predictive significance of ErbB-2 breast tumor levels measured by enzyme immunoassay. J. Clin. Oncol. **2001**, *19*, 645–656.

45. Fulton, R.J.; McDade, R.L.; Smith, P.L.; et al. Advanced multiplexed analysis with the FlowMetrix™ system. Clin. Chem. **1997**, *43*, 1749–1756.

46. Chen, R.; Lowe, L.; Wilson, J.D.; et al. Simultaneous quantification of six human cytokines in a single sample using microparticle-based flow cytometric technology. Clin. Chem. **1999**, *45*, 1693–1694.

7
Validation of Biomarkers as Surrogates for Clinical Endpoints

Marc Buyse
International Drug Development Institute, Cambridge, Massachusetts,
U.S.A.

Tony Vangeneugden and Luc Bijnens
Janssen Research Foundation, Beerse, Belgium

Didier Renard, Tomasz Burzykowski, Helena Geys, and
Geert Molenberghs
Limburgs Universitair Centrum, Diepenbeek, Belgium

I. INTRODUCTION

Biomarkers will become important in the clinic over the years to come, for several reasons. First, an increasing number of new drugs will have a well-defined mechanism of action at the molecular level, allowing drug developers to measure the effect of these drugs on the relevant biomarkers. Second, there will be increasing public pressure for new, promising drugs to be approved for marketing as rapidly as possible, and such approval will have to be based on biomarkers rather than on some long-term clinical endpoint. Finally, if the approval process is shortened, there will be a corresponding need for earlier detection of safety signals that could point to toxic problems with new drugs. It is a safe bet, therefore, that the evaluation of tomorrow's drugs will be based primarily on biomarkers, rather than on the longer-term, harder clinical endpoints that have dominated the development of new drugs until now.

Yet, for biomarkers to be acceptable surrogates for clinical endpoints, a number of conditions must be fulfilled. In this chapter, we review these conditions and we discuss some statistical methods that are useful to address the problem of surrogate marker validation. Much of the work laid out here is still in progress. The statistical approach proposed has been developed using data from a range of clinically diverse situations, including age-related macular degeneration [1–3], cardiovascular disease [2], advanced ovarian cancer [3,4], chronic schizophrenia [5,6], and advanced colorectal cancer [1,4,7,8]. It is currently being validated in other situations, including advanced prostate cancer, advanced breast cancer, early colorectal cancer, early breast cancer, and autoimmune deficiency syndrome (AIDS).

In this chapter, we concentrate on one clinical situation, the hormonal treatment of advanced (metastatic) prostate cancer, to illustrate the statistical methods used for, and the difficulties encountered in, the validation of a biomarker (the prostate-specific antigen, PSA, measured over time) as a surrogate for a clinical endpoint (the patient's death). We will avoid, insofar as possible, technical developments that have been published in full detail elsewhere [1–8]. Although some of our observations are specific to this situation, many of our conclusions are of general relevance to the validation of biomarkers as surrogates for clinical endpoints.

II. CONCEPTUAL FOUNDATION

A. Statistical Definitions and Models

Let us first introduce the problem in general terms, and define some notations that will be used throughout this chapter. We are interested in the effect of some experimental treatment on a clinical or "true" endpoint of interest, as well as on a biomarker that could potentially be used as a "surrogate" endpoint (Fig. 1). In general, the experimental treatment is compared to an appropriate control group in randomized clinical trials.

Statistically, interest will focus on the following parameters (Fig. 1): the effect of the experimental treatment upon the biomarker, called α; the effect of the experimental treatment upon the clinical endpoint, called β; and the effect of the surrogate biomarker on the clinical endpoint, called γ. It will be useful to denote the randomized treatment by Z, the potential surrogate biomarker by S, and the true clinical endpoint by T. Strictly speaking, the biomarker can be used as a surrogate for the clinical endpoint for the purposes of evaluating the experimental treatment if, and only if, a treatment effect on S ($\alpha \neq 0$) predicts a treatment effect on T ($\beta \neq 0$), and no treatment effect on S ($\alpha = 0$) predicts no treatment effect on T ($\beta = 0$). This view of surrogacy, which is rooted in the paradigm of hypothesis testing, had led to a formal statistical definition of surrogacy, but not to useful validation criteria [9–12]. Alternatively, the biomarker can be used as a surrogate for the clinical endpoint for the purposes of evaluating the experimental treatment if, and only if, the estimated treatment

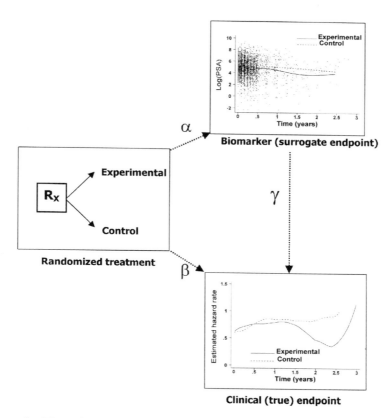

Figure 1 The validation of a biomarker (or intermediate endpoint) as a surrogate for a clinical endpoint (or true endpoint) with respect to the effect of a randomized treatment involves estimating parameters α, β, and γ. In advanced prostate cancer, the biomarker could be the level of PSA over time, and the clinical endpoint could be the time to death. Shown are individual PSA values and their mean over time by treatment group, and the hazard rate over time by treatment group.

effect on S (parameter α) can be used to predict the treatment effect on T (parameter β) with sufficient accuracy [1,3]. This view of surrogacy, which is rooted in the paradigms of estimation and prediction, will be adopted in our analyses of the data in advanced prostate cancer.

Let us first assume the simple, but rare, situation in which the biomarker S and the clinical endpoint T have a bivariate standardized normal distribution. The bivariate normal distribution has been extensively studied, and the statistical techniques required in this situation are straightforward. In reality, the situations will be more complex and will call for less standard models, but the underlying ideas will remain unchanged. If we had data from a single randomized clinical trial with n subjects, the relationships between Z, S, and T could be modeled through simple linear regressions:

$$S_i = \mu_S + \alpha Z_i + \varepsilon_{Si} \tag{1}$$

$$T_i = \mu_T + \beta Z_i + \varepsilon_{Ti} \tag{2}$$

$$T_i = \mu + \gamma S_i + \varepsilon_i \tag{3}$$

where μ_S, μ_T, and μ are intercepts; α, β, and γ are the slopes of the regression lines, and also the parameters of interest (Fig. 1); and ε_{Si}, ε_{Ti}, and ε_i are normally distributed error terms. The dependence of T upon both Z and S could be modeled through a multiple linear regression:

$$T_i = \mu' + \beta_S Z_i + \gamma_Z S_i + \varepsilon_i' \tag{4}$$

If we had data from several trials, the relationships between Z, S, and T would become:

$$S_{ij} = \mu_{Si} + \alpha_i Z_{ij} + \varepsilon_{Sij} \tag{1'}$$

$$T_{ij} = \mu_{Ti} + \beta_i Z_{ij} + \varepsilon_{Tij} \tag{2'}$$

with notations analogous to those used above, the subscript i now referring to trial and the subscript j to individual patients. In the most general case, a linear mixed model approach could be used, where the intercepts μ_{Si} and μ_{Ti}, as well as the slopes α_i and β_i, can be decomposed into fixed and random components [13]. We shall need these models to discuss validation criteria.

B. Types of Biomarkers and Endpoints

Statistically speaking, the biomarker and the clinical endpoint are realizations of random variables. Interest focuses on the joint distribution of these variables, which was assumed bivariate normal in the preceding models. This is, however, seldom the case, because the biomarker and/or the clinical endpoint is often a realization of nonnormally distributed random variables, which can be:

> *Binary* (dichotomous): biomarker value below or above a certain threshold (e.g., CD4+ counts over 500/mm^3) or clinical "success" (e.g., tumor shrinkage)
> *Categorical* (polychotomous): biomarker value falling in successive classes (e.g., cholesterol levels <200 mg/dL, 200–299 mg/dL, 300+ mg/dL) or

clinical response (e.g., complete response, partial response, stable disease, progressive disease)

Continuous (normally distributed): biomarker (e.g., log PSA level) or clinical measurement (e.g., diastolic blood pressure)

Censored continuous: time to biomarker below or above a certain threshold (e.g., time to undetectable viral load) or time to clinical event (e.g., time to cardiovascular death)

Longitudinal (repeated measures): biomarker (e.g., CD4+ counts over time) or clinical outcome (e.g., blood pressure over time)

Multivariate longitudinal: several biomarkers (e.g., CD4+ and viral load over time) or several clinical measurements (e.g., dimensions of quality of life over time)

The models used to validate a biomarker as a surrogate for a clinical endpoint will depend on the type of variables observed in the problem at hand. In the example below, we will illustrate this by analyzing the same data in three different ways. The clinical endpoint will be survival in all cases, but the biomarker will consist, respectively, of PSA response (binary variable), time to PSA progression (censored continuous variable), and the PSA pattern over time (longitudinal).

C. Types of Data

To validate the use of biomarkers as surrogates for clinical endpoints, the following information must be available on some series of patients:

Surrogate biomarker or endpoint: most commonly a vector of repeated measurements of the biomarker during the patient's treatment course or follow-up thereafter

Clinical endpoint: most commonly a time (possibly censored) to the clinical event of interest

Treatment: a categorical variable indicating what treatment the patient received (often through randomization)

Unit of analysis: typically a categorical variable indicating the "unit" in which the patient was treated (physician, center, country, trial, meta-analysis, or any other unit defining groups of patients in whom the effect of treatment can meaningfully be estimated)

III. STATISTICAL CRITERIA FOR SURROGACY

The purpose of this section is to provide a brief overview of various statistical ideas that have been proposed for the validation of markers as surrogates for clinical endpoints. In the next section, we will show through an actual example that some of these ideas lead to useful operational criteria.

A.   Measures of Association Between the Biomarker and the Clinical Endpoint

Several authors have argued that if a biomarker is to serve as a surrogate for a clinical endpoint, there should be a causal relationship between them [14,15]. If there was a causal pathway from the surrogate marker to the clinical endpoint, then any change in the marker (e.g., as a result of treatment) would translate into a corresponding change in the clinical endpoint. Causality, unfortunately, cannot be tested, and the statistical criteria developed to validate a surrogate marker provide only indirect evidence about the causality of the relationship between the marker and the endpoint.

A first source of evidence is provided by the association, *at the level of the individual patient*, between the marker and the clinical endpoint. One would expect a good surrogate marker to have a strong association with the clinical endpoint at the individual level, reflecting some biological pathway from the biomarker to the clinical endpoint. In that case, the biomarker could be a plausible surrogate on biological grounds, since the clinical endpoint would be largely determined by the biomarker regardless of any treatment effect. This reasoning, although intuitively appealing, has, however, been shown to be potentially misleading, for a good correlate is not automatically a good surrogate [15]. Another source of evidence is needed to quantify the association, *at the level of a trial*, between the effects of a treatment on the marker and on the clinical endpoint. The distinction between these two levels of evidence is essential, but has sometimes been missed in attempts to validate surrogate markers in the past [16]. We return to the trial-level association below.

The individual-level measure of association between the biomarker and the clinical endpoint could be provided by parameter γ_Z in Eq. (4), the slope of the linear regression line between S and T (adjusted for Z), or on a closely related parameter, the squared correlation between S and T (adjusted for Z), which has a more general and intuitive interpretation. The squared correlation (or coefficient of determination) represents the proportion of variance of the clinical endpoint that is explained by the variance of the biomarker, after adjusting for any difference due to treatment. We denote this coefficient $R^2_{individual}$ to stress that it characterizes the association between the biomarker and the clinical endpoint in individual patients. Just as in linear regression, we will require $R^2_{individual}$ to be large (close to 1) before we claim that there is a strong association between the biomarker and the clinical endpoint.

For biomarkers and clinical endpoints that are not normally distributed, other measures of association will be used, as will be shown in the analyses below, but the basic idea of a strong association between the biomarker and the clinical endpoint will carry over.

B.   Explanation of Clinical Effects from Surrogate Effects

Prentice proposed to define a surrogate marker as "a response variable for which a test of the null hypothesis of no relationship to the treatment groups under comparison is also a valid test of the corresponding null hypothesis based on the true endpoint" [9]. As such, this definition is of limited value since direct verification that a triplet {treatment; surrogate biomarker; clinical endpoint} fulfills the definition would require a large number of experiments to be available with information on the triplet. Even if many experiments were available, the equivalence of the statistical tests for the effect of treatment upon the clinical endpoint and the biomarker might not be seen in all of them because of chance fluctuations and/or lack of statistical power. Operational criteria are therefore needed to check if the definition is fulfilled. Prentice proposed four operational criteria:

Treatment must have a significant effect on the biomarker [$\alpha \neq 0$ in Eq. (1)].

Treatment must have a significant effect on the clinical endpoint [$\beta \neq 0$ in Eq. (2)].

The biomarker must have a significant effect on the clinical endpoint [$\gamma \neq 0$ in Eq. (3)].

The *full* effect of treatment on the clinical endpoint must be captured by the biomarker [$\beta_S = 0$ in Eq. (4)].

Even though the prentice criteria were of key importance to help formalize validation approaches, a number of conceptual problems were identified with them. Indeed, it can be shown that Prentice's operational criteria are equivalent to his definition only in the case of binary variables [1]. Moreover, the operational criterion of *full* capture raises a conceptual difficulty in that it requires the statistical test for treatment effect on the true endpoint to be *non*significant after adjustment for the surrogate [11]. Hence this criterion is useful only to reject a poor surrogate biomarker, when the statistical test for treatment effect on the true endpoint remains statistically significant after adjustment for the surrogate. An example of such a situation is given by the effects of zidovudine on clinical endpoints in human-immunodeficiency-virus-positive subjects, which remain significant after CD4+ lymphocytes are taken into account [17,18].

The fourth Prentice criterion cannot be used as such to validate a good surrogate marker, for failing to reject the null hypothesis may be due merely to lack of power. Freedman et al. therefore suggested focusing attention on the proportion of the treatment effect captured by the surrogate, or "proportion explained" [11,19]. In this spirit, a good surrogate is one that explains a large proportion of that effect. Numerically, the proportion explained can be estimated as the ratio $(\beta - \beta_S)/\beta$ from Eqs. (2) and (4). Calculation of its confidence limits requires estimation of the covariance between β and β_S. Several authors have shown that there are fundamental difficulties with the proportion explained, and have proposed alternative approaches [1,12,20].

C. Prediction of Clinical Effects from Surrogate Effects

The reason for using surrogate markers (or surrogate endpoints) is to be able to predict the effect of treatment on the clinical endpoint, having observed its effect on the surrogate marker. This led to consideration of the ratio of the effect of treatment on the clinical endpoint to that on the surrogate marker, or "relative effect" [1]. Numerically, the relative effect can be estimated as the ratio β/α from Eqs. (1) and (2). Calculation of its confidence limits requires estimation of the covariance between β and α. Note that the relative effect depends on the scales chosen to measure S and T. If the relative effect is estimated precisely, then the predicted effect upon the clinical endpoint will in turn be precise enough to be useful. Such a situation requires large numbers of observations that are typically available in large clinical trials, or in meta-analyses of several clinical trials. When meta-analytical data are available, it is also possible to test the assumption implicit in the estimation of the relative effect, i.e., that the treatment effects on the clinical endpoint are proportional to the treatment effects on the surrogate biomarker.

D. Measures of Association Between Treatment Effects

If data are available from multiple sources, for instance if several clinical trials have been performed on the same therapy, it will be possible to estimate the treatment *effects* upon the marker and upon the clinical endpoint in each of these trials [3,21,22]. These treatment effects were denoted α_i and β_i in Eqs. (1') and (2'). We focus here on the squared correlation (or coefficient of determination) between these treatment effects, which represents the proportion of variance of the treatment effect on the clinical endpoint that is explained by the variance of the treatment effect on the biomarker. We denote this coefficient R^2_{trial} to stress that it characterizes the association between the effects of treatment on the biomarker and on the clinical endpoint in the various trials available. Here again, we will require R^2_{trial} to be large (close to 1) before we claim that there is a strong association between the effects on the biomarker and on the clinical endpoint.

IV. EXAMPLE: PSA IN ADVANCED PROSTATE CANCER

A. The Two Liarozole Trials

We illustrate the statistical approach based on the individual-level and trial-level associations using two trials in patients with advanced (metastatic) prostate cancer. These trials compared oral liarozole, an experimental retinoic acid metabolism-blocking agent developed by the Janssen Research Foundation, with two antiandrogenic drugs: cyproterone acetate (CPA) in the first trial and flutamide in the second. In both trials, patients were in relapse after first-line endocrine therapy [23]. The trials accrued 312 and 284 patients, respectively. Each trial was multinational and multicentric. Since our analyses require the estimation of the effect of treatment in multiple trials or other meaningful groups of patients, we grouped the patients by trial and by country. This allowed us to define 19 groups containing between four and 69 patients per group.

The primary endpoint of the trials was overall survival from the start of treatment. Assessments were undertaken before the start of treatment, at 2 weeks, monthly for 6 months, at 3-month intervals until the second year, and at 6-month intervals until treatment discontinuation or death. The assessments included measurement of the prostate-specific antigen (PSA) level. PSA is a glycoprotein that is found almost exclusively in normal and neoplastic prostate cells. Changes in PSA often antedate changes in bone scan, and they have been used as an indicator of response in patients with androgen-independent prostate cancer [24–26].

We consider, successively, PSA response, time to PSA progression (TPP), and the full longitudinal PSA profile of each patient as potential surrogates for survival in this disease [27].

B. PSA Response as Surrogate for Survival

The best PSA outcome was determined for each patient, and hierarchically ordered as [28]:

> *Complete response* (CR) if the PSA level was at least 20 ng/mL at baseline, returned to normal (< 4 ng/mL) at any time, and remained normal for at least 28 days
> *Partial response* (PR) if the PSA level was at least 20 ng/mL at baseline, decreased by at least 50% from the baseline level, and remained under 50% of the baseline level for at least 28 days

No change (NC) if the PSA level was at least 20 ng/mL at baseline, and fluctuated between 50% below and 50% above the baseline level for at least 28 days

Progressive disease (PD) if no other response category applied, and if PSA was at least equal to 10 ng/mL

Not evaluable (NE), if none of the above applied

A patient was defined as having a PSA response if his best PSA outcome was either PR or CR. Hence the biomarker is binary here, and the clinical endpoint is a (possibly censored) survival time.

At the individual level, PSA response was a very strong predictor of survival (Fig. 2a). Because PSA response is binary and survival is censored, the normal theory coefficient of determination (R^2) discussed earlier does not apply, and another measure of association between PSA response and survival is needed. One way to express the impact of PSA response on survival is as follows [8]: consider the odds of surviving beyond time t for PSA responders and for nonresponders; the ratio of these odds is a survival odds ratio. Although the odds of surviving beyond time t decrease in time for both responders and nonresponders, in our model the *ratio* of these odds is assumed constant. This survival odds ratio is equal to 5.5 (95% confidence interval $= 2.7–8.2$), which means that at any point in time the odds of surviving beyond that time are more than five times higher for patients with a PSA response as compared to patients without such a response. The strong prognostic impact of PSA response can be explained in at least three plausible ways:

PSA response and survival are largely determined by a common set of prognostic factors, so that patients who are likely to have a response are also those who are potentially long survivors.

Patients who survive a long time are more likely to have a PSA response because of length-biased sampling [29].

There is a true causal relationship between the achievement of a PSA response and a prolongation of survival.

The first and second explanations are amenable, at least in part, to statistical investigations, the first through adjustments of the comparison of responders and nonresponders for all known prognostic factors, and the second through a landmark analysis [30]. When these investigations fail to explain a large portion of the prognostic impact of PSA response, then there is indirect evidence that PSA response truly results in a survival improvement [7].

At the group level, the effects of liarozole on PSA response and on survival were poorly correlated, with a coefficient of determination $R^2_{trial} = 0.05$ (standard error $= 0.13$) (Fig. 2b).

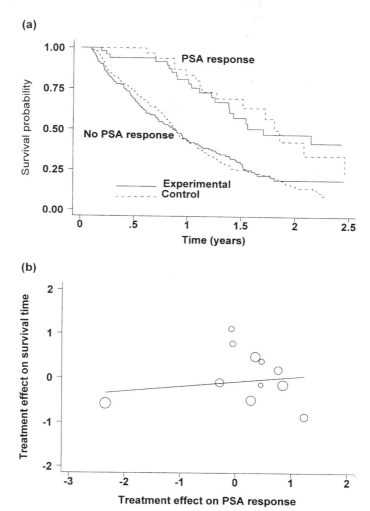

Figure 2 (a) The survival of patients with a PSA response differs substantially from that of patients without a PSA response. At any point in time the odds of surviving beyond that time are more than five times higher for patients with a PSA response as compared to patients without such a response (see text). (b) The treatment effects on survival and on PSA response show no correlation in advanced prostate cancer ($R^2_{trial} = 0.05$). Each circle shows treatment effects estimated in one of the countries in which the trials were conducted. (The size of the circle is proportional to the number of patients.)

There was no overall significant benefit of liarozole over control for either response or survival: the PSA response rate was 16% and 11%, respectively, for liarozole and control ($p = 0.11$), while median survival was 11.3 and 10.9 months, respectively, for liarozole and control ($p = 0.71$).

C. Time to PSA Progression as Surrogate for Survival

The time to PSA progression (TPP) was determined on the basis of a moving average of three consecutive values of PSA. Progression was defined as an increase in PSA equal to, or larger than, 50% above the lowest prior moving average. This increase had to be either the last determination in the patient's follow-up, or maintained for at least 28 days.

At the individual level, PSA progression occurred much earlier than the patients' death. PSA progression occurred within 6 months for half of the patients, while about half of the patients were still alive at 1 year (Fig. 3a). Here again, because TPP and survival may both be censored, the normal theory coefficient of determination (R^2) discussed earlier does not apply, and a possible measure of association between TPP and survival is a generalization of that proposed above [4]: consider the odds of surviving beyond time t for patients who have not yet had a PSA progression, and for those who have; the ratio of these odds is a survival odds ratio. Although the odds of surviving beyond time t decrease in time for both patients with and without PSA progression, in our model the *ratio* of these odds is assumed constant.

This odds ratio is equal to 6.3 (95% confidence interval $= 4.4-8.2$), which means that at any point in time the odds of surviving beyond that time are more than six times higher for patients who have not yet had a PSA progression as compared to patients who have already had such a progression. Thus, here again, there is a strong individual-level association between TPP and survival.

At the group level, the effects of liarozole on TPP and on survival were poorly correlated, with a coefficient of determination $R^2_{trial} = 0.22$ (standard error $= 0.18$) (Fig. 3b). There was a significant benefit of liarozole over control in terms of time to PSA progression, with a median time of 4.9 months for liarozole and 3.7 months for control ($p = 0.001$).

D. Longitudinal Measurements of PSA as Surrogate for Survival

Since PSA levels were measured repeatedly over time, it seems natural to make use of all these measurements, rather than to define a single PSA response or time to PSA progression for each patient. The statistical models required to take the longitudinal nature of the measurements into account are more complex, and the analyses potentially more sensitive to model assumptions, than for singly measured

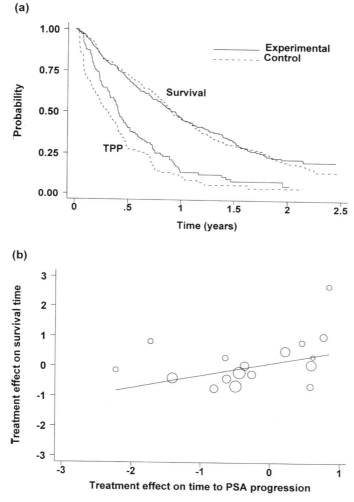

Figure 3   (a) PSA progression is a strong predictor of death in advanced prostate cancer. At any point in time the odds of surviving beyond that time are more than six times higher for patients who have not yet had a PSA progression as compared to patients who have already had such a progression (see text). (b) The treatment effects on survival and on time to PSA progression show very little correlation in advanced prostate cancer ($R^2_{trial} = 0.22$).

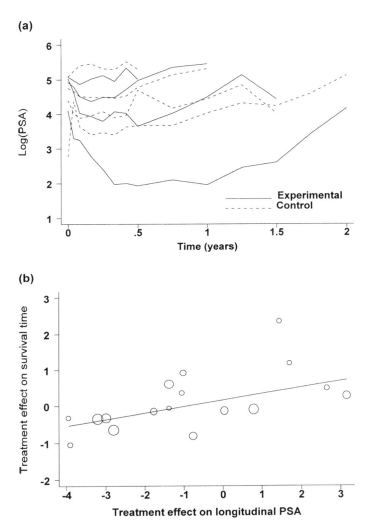

Figure 4 (a) The mean PSA profiles for cohorts of patients with similar follow-up times show a tendency for PSA to go down initially (PSA response), and to come up again after a while (PSA progression). The longitudinal PSA profiles are strongly correlated with the hazard of death ($R^2_{individual} > 0.84$ at any point in time). (b) The treatment effects on survival and on longitudinal PSA show a weak correlation in advanced prostate cancer ($R^2_{trial} = 0.42$).

endpoints. Such models have been used extensively to study the relationship between CD4 lymphocytes and survival in patients with AIDS and AIDS-related complex [31–35].

In our example, the mean PSA levels over time shown in the upper-right-hand panel of Fig. 1 are not fully informative, because these means were not calculated on the same patients over time. Indeed, patients who had a PSA progression left the study, and no longer contributed to the mean PSA after that time point, thus creating a selection bias in the calculation of the mean. A more informative way of looking at mean PSA levels over time is to consider cohorts of patients defined by the time they leave the study (for any reason). Figure 4a shows four such cohorts, split by treatment group: patients leaving the study within 6 months, between 6 and 12 months, between 12 and 18 months, and between 18 and 24 months (PSA data became too scarce to calculate meaningful means after 24 months). The patterns exhibited by these cohort-specific means show a tendency for PSA to go down initially (PSA response), and to come up again after a while (PSA progression).

At the individual level, the PSA longitudinal process was correlated with the hazard rate, which is the risk of dying at a certain time for a patient who has survived up until that time. The coefficient of determination between the PSA process and the hazard rate ($R^2_{individual}$) is here a function of time that cannot be easily summarized into a single measure [5]. Suffice to say that $R^2_{individual}$ was greater than 0.84 at all times to indicate that there was again a strong association, at the individual patient level, between the evolution of PSA and the hazard of dying.

At the group level, the effects of liarozole on longitudinal PSA and on survival were moderately correlated, with a coefficient of determination $R^2_{trial} = 0.45$ (standard error = 0.18) (Fig. 4b). There was a significant benefit of liarozole in terms of longitudinal PSA ($p = 0.01$); in other words, the profiles shown on Fig. 4a were significantly different between liarozole and control.

V. DISCUSSION

We have illustrated, through an actual example, statistical approaches that may be useful to study the complex relationships between a biomarker, a clinical endpoint, and the effects of a treatment on both the biomarker and the clinical endpoint. Our analyses emphasize the importance of distinguishing between two types of association: one between the biomarker and the clinical endpoint at the individual level, the other between the effects of treatment on the biomarker and on the clinical endpoint at the trial (or group) level. Since only two trials were available for our analyses, we considered country in each trial as the grouping unit of interest. Table 1 summarizes the measures of association between survival and,

Table 1 Individual-Level and Trial-Level Measures of Association Between PSA and Survival in Advanced Prostate Cancer Treated with Either Liarozole or Control[a]

	Individual-level association between PSA and survival [95% confidence interval]	Trial-level association between treatment effects on PSA and survival [standard error]
PSA response	Survival odds ratio = 5.5 (2.7–8.2)	$R^2_{trial} = 0.05\,(0.13)$
Time to PSA progression	Survival odds ratio = 6.3 (4.4–8.2)	$R^2_{trial} = 0.22\,(0.18)$
Longitudinal PSA	Coefficient of determination $R^2(t) > 0.84$ at all times t	$R^2_{trial} = 0.45\,(0.18)$

[a] The individual-level measures show strong associations between PSA and survival, but the trial-level measures show weak associations between the treatment effects on PSA and survival, making PSA a poor surrogate for survival (odds ratio: see text; R^2 = coefficient of determination).

successively, response to PSA, time to PSA progression, and longitudinal PSA (rows in Table 1). It appears clearly that PSA does not qualify as an acceptable surrogate, regardless of how it is analyzed, in spite of its strong associations with survival at the individual level (second column of Table 1). The associations between treatment effects at the trial level are all low (third column of Table 1). Even when the full PSA pattern is taken into account in a longitudinal analysis, R^2_{trial} is still too low to permit reliable prediction of the effect of treatment on the clinical endpoint, having observed the effect of treatment on the biomarker.

It is also clear from Table 1 that the trial-level associations are estimated rather imprecisely, because of the relatively small number of units (countries) available to estimate treatment effects. In general, the individual-level associations can be estimated far more precisely, because of the large number of patients available [1–8].

It should be noted that the methodology we propose is exploratory in nature, and does not purport to classify a biomarker as a "valid" or "invalid" surrogate for a clinical endpoint—although if both $R^2_{individual}$ and R^2_{trial} were close to 1, we would be in a position to claim the surrogate to be acceptable. Indeed, in such a case, the surrogate would be strongly associated to the clinical endpoint, and any *change* in the surrogate would also translate into a corresponding (and predictable) change in the clinical endpoint. However, caution would still be in order, for neither of these statistical associations would prove a causal impact of the biomarker on the clinical endpoint. Moreover, the trial-level association would have been established only for the treatment comparison at hand, and could be quite different for some new treatment having a different mode of action.

The validation of a biomarker as a surrogate for a clinical endpoint is no easy task. Many authors have expressed an exceedingly negative view on this problem. Theoretical criticisms have borne on problems with overly strict definitions of surrogacy [12,15,20], the validation criteria proposed by Prentice [12,36], the proportion explained [12,20], computation and modeling difficulties [37], and the meta-analytic approach [38]. On the practical side, some supposed surrogates have dramatically failed to predict clinical outcomes [39]. The approval of the antiarrhythmic drugs flecanaide and encanaide, based on their controlling arrhythmias rather than long-term mortality, will long continue to haunt the debates on whether surrogate endpoints can be used to approve new drugs [15,40]. It seems clear that few, if any, biomarkers will ever qualify as "valid" surrogates in a strict sense of the word. Even if we adopt the more liberal view advocated in this chapter, very few, if any, biomarkers will have large enough values of R^2 to qualify as "acceptable" surrogates [41]. In addition, surrogates that are observed very early on in the course of the disease are the most interesting ones, but also those least likely to predict distant clinical endpoints with any acceptable accuracy. In spite of all difficulties, we believe that the search for surrogates should not be abandoned, for the gains might be too important in terms of patients and/or time. For some endpoints, such as delayed toxicities to experimental treatments, the use of surrogates is simply inescapable. In addition, even if biomarkers always turned out to be poor surrogates, it could still be useful to quantify their relationships to the clinical endpoints of interest, because valuable knowledge might well be derived in the process.

A final word on the need for data. The methods presented here require data from several (possibly many) randomized trials to be available. Access to data from randomized trials is difficult, especially for phase III trials carried out by pharmaceutical companies seeking registration of new drugs. We contend that the only way to seriously search for valid surrogate biomarkers is to make these data fully accessible for statistical analysis and public scrutiny. Once new drugs are approved, individual patient data from randomized clinical trials upon which the approval was based should be made publicly accessible, as are data from some cooperative groups (the AIDS Clinical Trials Group, for instance). Further analyses of such data in clinical situations of interest may illuminate issues related to surrogate endpoints that, in the absence of detailed statistical analyses, would have remained controversial at best, and ignored at worst.

ACKNOWLEDGMENTS

The authors thank the Janssen Research Foundation for permission to use data from two clinical trials testing liarozole in patients with advanced prostrate cancer.

REFERENCES

1. Buyse, M.; Molenberghs, G. Criteria for the validation of surrogate end-points in randomized experiments. Biometrics **1998**, *54*, 1014–1029.
2. Molenberghs, G.; Geys, H.; Buyse, M. Evaluation of surrogate end-points in randomized experiments with mixed discrete and continuous outcomes. Stat. Med. **2001**, *20*, 3023–3038.
3. Buyse, M.; Molenberghs, G.; Burzykowski, T.; Renard, D.; Geys, H. The validation of surrogate endpoints in meta-analyses of randomized experiments. Biostatistics **2000**, *1*, 49–68.
4. Burzykowski, T.; Molenberghs, G.; Buyse, M.; Geys, H.; Renard, D. Validation of surrogate endpoints in multiple randomized clinical trials with failure-time endpoints. Appl. Stat. **2001**, *50*, 405–422.
5. Renard, D.; Geys, H.; Molenberghs, G.; Burzykowski, T.; Buyse, M. Validation of surrogate endpoints in multiple randomized clinical trials with discrete outcomes. Biom. J. **2002**, *44*, 921–935.
6. Alonso, A.; Geys, H.; Molenberghs, G.; Vangeneugden, T. Investigating the criterion validity of psychiatric symptom scales using surrogate marker validation methodology. J. Biopharmaceutical Statist., **2002**, *12*, 161–178.
7. Buyse, M.; Thirion, P.; Carlson, R.W.; Burzykowski, T.; Molenberghs, G.; Piedbois, P., for the Meta-Analysis Group in Cancer. Relation between tumour response to first-line chemotherapy and survival in advanced colorectal cancer: a meta-analysis. Lancet **2000**, *356*, 373–378.
8. Burzykowski, T.; Molenberghs, G.; Buyse, M.; Geys, H.; Renard, D. The validation of surrogate endpoints using data from randomized clinical trials: a case study in advanced colorectal cancer. J. Royal Statist. Soc. A, **2003** (in press).
9. Prentice, R.L. Surrogate endpoints in clinical trials: definitions and operational criteria. Stat. Med. **1989**, *8*, 431–440.
10. Schatzkin, A.; Freedman, L.S.; Schiffman, M.H.; Dawsey, S.M. Validation of intermediate end points in cancer research. J. Natl Cancer Inst. **1990**, *82*, 1746–1752.
11. Freedman, L.S.; Graubard, B.I.; Schatzkin, A. Statistical validation of intermediate endpoints for chronic diseases. Stat. Med. **1992**, *11*, 167–178.
12. Molenberghs, G.; Buyse, M.; Geys, H.; Renard, D.; Burzykowski, T. Statistical challenges in the evaluation of surrogate endpoints in randomized trials. Control. Clin. Trials., **2002**, *23*, 607–625.
13. Verbeke, G.; Molenberghs, G. *Linear Mixed Models for Longitudinal Data*; Springer Series in Statistics; Springer: New York, 2000.
14. Lagakos, S.W.; Hoth, D.F. Surrogate markers in AIDS: where are we? Where are we going? Ann. Intern. Med. **1992**, *116*, 599–601.
15. Fleming, T.R.; DeMets, D.L. Surrogate end points in clinical trials: are we being misled? Ann. Intern. Med. **1996**, *125*, 605–613.
16. Jacobson, M.A.; Bacchetti, P.; Kolokathis, A.; et al. Surrogate markers for survival in patients with AIDS and AIDS related complex treated with zidovudine. Br. Med. J. **1991**, *302*, 73–78.

17. Lin, D.Y.; Fischl, M.A.; Schoenfeld, D.A. Evaluating the role of CD4-lymphocyte change as a surrogate endpoint in HIV clinical trials. Stat. Med. **1993**, *12*, 835–842.
18. Choi, S.; Lagakos, S.; Schooley, R.T.; Volberding, P.A. CD4$^+$ lymphocytes are an incomplete surrogate marker for clinical progression in persons with asymptomatic HIV infection taking zidovudine. Ann. Intern. Med. **1993**, *118*, 674–680.
19. Lin, D.Y.; Fleming, T.R.; De Gruttola, V. Estimating the proportion of treatment effect explained by a surrogate marker. Stat. Med. **1997**, *16*, 1515–1527.
20. Flandre, P.; Saidi, Y. Letter to the editor: estimating the proportion of treatment effect explained by a surrogate marker. Stat. Med. **1999**, *18*, 107–115.
21. A'Hern, R.P.; Ebbs, S.R.; Baum, M.B. Does chemotherapy improve survival in advanced breast cancer? A statistical overview. Br. J. Cancer **1988**, *57*, 615–618.
22. Daniels, M.J.; Hughes, M.D. Meta-analysis for the evaluation of potential surrogate markers. Stat. Med. **1997**, *16*, 1515–1527.
23. Debruyne, F.J.M.; Murray, R.; Fradet, Y.; Johansson, J.E.; Tyrrell, C.; Boccardo, F.; Denis, L.; Marberger, J.M.; Brune, D.; Rassweiler, J.; Vangeneugden, T.; Bruynseels, J.; Janssens, M.; de Porre, P., for the Liarozole Study group. Liarozole—a novel treatment approach for advanced prostate cancer: results of a large randomized trial versus cyproterone acetate. Urology **1998**, *52*, 72–81 .
24. Sridhara, R.; Eisenberger, M.A.; Sinibaldi, V.J.; et al. Evaluation of prostate-specific antigen as a surrogate marker for response of hormone-refractory prostate cancer to suramin therapy. J. Clin. Oncol. **1995**, *13*, 2944–2953.
25. Smith, D.C.; Dunn, R.L.; Stawderman, M.S.; et al. Change in serum prostate-specific antigen as a marker of response to cytotoxic therapy for hormone-refractory prostate cancer. J. Clin. Oncol. **1998**, *16*, 1835–1843.
26. Kelly, W.K.; Scher, H.I.; Mazumdar, M.; et al. Prostate-specific antigen as a measure of disease outcome in metastatic hormone-refractory prostate cancer. J. Clin. Oncol. **1993**, *11*, 607–615.
27. Scher, H.I.; Kelly, W.K.; Zhang, Z.F.; et al. Post-therapy serum prostate-specific antigen level and survival in patients with androgen-independent prostate cancer. J. Natl Cancer Inst. **1999**, *91*, 244–251.
28. Bubley, G.J.; Carducci, M.; Dahut, W.; Dawson, N.; Daliani, D.; Eisenberger, M.; Fidd, W.D.; Freidlin, B.; Halabi, S.; Hudes, G.; Hussain, M.; Kaplan, R.; Myers, C.; Oh, W.; Petrylak, D.P.; Reed, E.; Roth, B.; Sartor, O.; Scher, H.; Simons, J.; Sinibaldi, V.; Small, E.J.; Smith, M.R.; Trump, D.L.; Vollmer, R.; Wilding, G. Eligibility and response guidelines for phase II clinical trials in androgen-independent prostate cancer: recommendations from the prostate-specific antigen working group. J. Clin. Oncol. **1999**, *17*, 3461–3467.
29. Buyse, M.; Piedbois, P. On the relationship between response to treatment and survival. Stat. Med. **1996**, *15*, 2797–2812.
30. Anderson, J.R.; Cain, K.C.; Gelber, R.D. Analysis of survival by tumor response. J. Clin. Oncol. **1983**, *1*, 710–719.

31. De Gruttola, V.; Wulfsohn, M.; Fischl, M.A.; Tsiatis, A. Modelling the relationship between survival and CD4 lymphocytes in patients with AIDS and AIDS-related complex. J. AIDS **1993**, *6*, 359–365.

32. De Gruttola, V.; Tu, X.M. Modelling progression of CD-4 lymphocyte count and its relationship to survival time. Biometrics **1995**, *50*, 1003–1014.

33. Diagnostic and therapeutic technology assessment (DATTA): surrogate markers of progressive HIV disease. J. Am. Med. Assoc. **1992**, *267*, 2948–2952.

34. Ellenberg, S.S. Surrogate endpoints in clinical trials: getting closer to identifying markers for survival in AIDS. Br. Med. J. **1991**, *302*, 63–64.

35. Machado, S.G.; Gail, M.H.; Ellenberg, S.S. On the use of laboratory markers as surrogates for clinical endpoints in the evaluation of treatment for HIV infection. J. AIDS **1990**, *3*, 1065–1073.

36. Begg, C.; Leung, D. On the use of surrogate endpoints in randomized trials. J. R. Stat. Soc. A **2000**, *163*, 26–27.

37. Tibaldi, F.S.; Abrahantes, J.C.; Molenberghs, G.; Renard, D.; Burzykowski, T.; Buyse, M.; Parmar, M.; Stijnen, T.; Wolfinger, R. Simplified hierarchical linear models for the evaluation of surrogate endpoints. J. Stat. Computation Simul. **2003** (in press).

38. Gail, M.H.; Pfeiffer, R.; van Houwelingen, H.C.; Carroll, R.J. On meta-analytic assessment of surrogate outcomes. Biostatistics **2000**, *1*, 231–246.

39. Temple, R.J. A regulatory authority's opinion about surrogate endpoints. In *Clinical Measurement in Drug Evaluation*; Nimmo, W., Ticker, G., Eds.; Chichester: Wiley, 1995.

40. The Cardiac Arrhythmia Suppression Trial (CAST) Investigators; Preliminary report: effect of encainide and flecainide on mortality in a randomized trial of arrhythmia suppression after myocardial infraction. N. Engl. J. Med. **1989**, *321*, 406–412.

41. Buyse, M.; Molenberghs, G.; Burzykowski, T.; Renard, D.; Geys, H. Statistical validation of surrogate endpoints: problems and proposals. Drug Inf. J. **2000**, *34*, 447–454.

8

Biomarkers for Pharmacokinetic/Pharmacodynamic Modeling and Clinical Trial Simulations

Wayne A. Colburn
MDS Pharma Services, Phoenix, Arizona, U.S.A.

I. INTRODUCTION

Biomarkers have begun to assume their place in the drug development process [1–6]. As has been discussed in some of the earlier chapters, when biomarkers are (1) grounded in mechanism-based disease and therapeutic intervention theory, (2) developed from discovery through preclinical assessments, and (3) measured with good laboratory practice (GLP)-like analytical methods, they have the potential to serve as a tool for early decision making and ultimately to become surrogate endpoints that predict clinical endpoints. But more to the point for this chapter, biomarkers have the potential to effectively lead drug development from drug-target rationale to discovery to preclinical development to clinical development to regulatory approval and labeling information via pharmacokinetic/pharmacodynamic (PK/PD) modeling and clinical trial simulations. Terms and definitions used in this chapter are those selected by the National Institutes of Health Biomarkers Definitions Working Group [7]. The term "clinical endpoint" is used rather than clinical outcome to avoid discussions centering on differences between clinical pharmacological and epidemiological perspectives.

PK/PD modeling and simulation can be effective tools to improve efficiency in the drug development process. PK/PD modeling can serve as a means to assess dose–concentration–effect relationships. PK/PD modeling and simulation can serve as methods to evaluate previously untested study designs, dose levels, and/or dosing regimens. PK/PD modeling and simulation can serve as tools for communication and education; for pharmacokineticists to more effectively communicate with other drug development experts; and to educate higher-level managers who are removed from the science and the medical realities of drug development. PK/PD modeling and simulation also is a tremendous educational tool for other stakeholders in the drug development and approval process such as practicing pharmacists and physicians, as well as regulators, health care managers, and patients.

Biochemical markers such as leukotrienes, chemokines, and cytokines, as well as clinical markers such as pulmonary function tests for asthma and chronic obstructive pulmonary disease; glucose, fructosamine, glycated albumin, HbA1c, and cytokines as well as retinal nephropathy, or peripheral neuropathy assessments for type 1 diabetes; or angiotensin I, angiotensin II, renin, aldosterone, and adrenocortical extract activity as well as electrocardiograms, blood pressure, and heart rate measures for hypertension, as well as clinical endpoints such as life or death; cure or failure or time to an event can be used as PD measures for PK/PD modeling. Plasma drug concentrations for drugs that need to be delivered to their site of action via the vascular space are biomarkers or surrogate endpoints for bioequivalence evaluation in our current regulatory paradigm. In a similar fashion, for PK/PD modeling and simulation, drug concentrations are simply the midway point in the dose–concentration–effect relationship. Drug concentrations in plasma are surrogates for drug concentrations in other tissues including the site(s) of action for both beneficial and adverse effects.

Biomarkers and surrogate endpoints have the potential to drive PK/PD modeling and simulation to new heights or, if not properly used, have the potential to limit the acceptance of PK/PD modeling and simulation in drug development and medical communication. Biomarkers can be precursors to later events such as clinical endpoints. Biomarkers have the potential to be measured more reproducibly and precisely than clinical endpoints... or maybe not. Achieving the true potential for biomarkers is up to the analytical chemists and the people in the drug development process who determine what methods will be used. Biomarker value is derived with respect to time and quality of the measurement (see Table 1). Biomarkers must be (1) available for measurement before the clinical endpoint presents itself, or (2) if the biomarker and clinical endpoint present at about the same time relative to disease progression, the biomarker must be measured in a more precise and reproducible way. In addition, biomarkers that are not grounded in sound theory, disease mechanism and/or mechanism-based therapeutic intervention can limit or totally destroy the potential power of otherwise science-based PK/PD modeling and simulation

Table 1 Using Biomarkers for Efficient Drug Development

Selection criteria
Time to onset of the biomarker must be short relative to time of onset of the clinical endpoint.
The analytical method for the biomarker must be at least as robust as for the clinical endpoint.
Predictive capability is mandatory for a surrogate endpoint.
Timing
Identify and evaluate safety and efficacy markers for decision making during preclinical R&D.
Validate biomarker assay methods during preclinical R&D.
Continue to evaluate and increase confidence in clinical application of the markers during phase I and IIa.
Expand validation criteria during clinical development.
Continue to evaluate and link clinical markers to clinical endpoints during phase IIb/III.

results. The remainder of this chapter will expand on these concepts to establish the use of biomarkers in PK/PD modeling and simulation as a central process to make drug development more efficient as well as more effective.

There are three corners to the triangle presented in Fig. 1 that set the conceptual foundation for PK/PD modeling and simulation: (1) mechanism-based models of disease, (2) mechanism-based therapeutic interventions, and (3) relationships between plasma drug concentrations and therapeutic and toxic responses. Functional genomics and proteomics provide support for the first pillar by identifying mechanism-based disease. Proteomics provides support for the second pillar by providing targets for mechanism-based therapeutic interventions. Finally, the collision between the pharmaceutical industry and the agencies that regulate drugs provides support for the third pillar with archives full of dose–concentration–effect relationship data that have been used for new drug approvals as well as generic drug and 505b2 approvals. More evidence for this view will be provided later to challenge those who have said, "Yeah, that's okay for some drugs, but what about drugs like central-nervous-system-active agents that have no concentration–effect relationship?" This perception is based on a lack of understanding of PK/PD principles and, in most cases, lack of effort to find potentially complex temporal relationships between drug concentrations in plasma and pharmacological effects at the site of action [8]. For drugs that exert their effects at sites other than the site of drug administration, plasma drug concentrations can be correlated both with drug concentrations at the site of action and, therefore, with observed effects. This will be discussed in more detail later.

The Conceptual Foundation
PK/PD Modeling

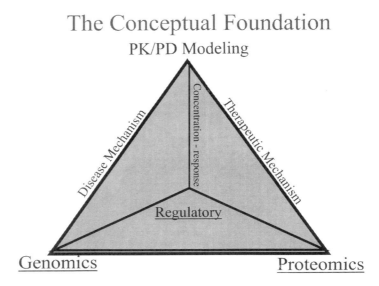

Figure 1 The conceptual foundation from which future PK/PD modeling and simulation will evolve. As functional genomics and proteomics provide the mechanistic basis for disease and therapeutic intervention, modeling and simulation will become even more beneficial for drug development.

II. PHARMACOKINETIC INPUT

Research articles, chapters, and books have been written on bioanalytical methods, pharmacokinetics, and their use as inputs for PK/PD modeling and will not be repeated here [9]. This chapter will discuss how this input, and issues associated with it, can influence the overall output from PK/PD modeling and simulation.

Clearly, the ability to create a predictive PK/PD model is easier if there is only one active moiety for PK input: parent drug *or* a single active metabolite. However, this is often not the case. Many drug substances are active and also have one or more active metabolites. If there is more than one PK input, such as parent plus an active metabolite or multiple active metabolites, the PK component of the PK/PD modeling becomes a multidimensional process. Multiple PK inputs obviously add complexity to the modeling exercise and increase the potential to have several nonunique solutions to a much more complicated PK/PD model. A total lack of information or limited information about the PK characteristics of the metabolites and/or the fractional contribution of each moiety to total in vivo pharmacological and toxicological activity can also add confusion to the modeling process. The bottom line is less confidence in the PK/PD model predictive capability and therefore a need to conduct more studies to bolster confidence in the output.

Another issue that needs to be considered is whether to use total or free drug concentrations in blood, serum, or plasma to establish the relevant drug concentrations at the site of action. In a linear drug-binding system, free and total drug concentrations can be used interchangeably to determine the amount of drug delivered to the site of action; free drug is simply a fraction of total drug across the range of plasma drug concentrations. However, even if plasma binding is linear, free concentrations near the site of action may not be the same as free concentrations in plasma. Extremely tight binding to a receptor or an enzyme can limit the equilibrium between free fraction in plasma and free fraction at the site of action. Conversion of the attached ligand to another entity before eliciting a response or before dissociation from the protein can result in a similar divergence of drug concentrations and effect. Free fraction can serve as a scaling function, but it can add to the difficulty in translating in vitro to in vivo results under the circumstances listed above.

A comprehensive PK profile provides the best input for PK/PD modeling and simulation, whether it is derived from a single dose or during a dosing interval at steady state [9]. During drug development, many clinical studies do not capture a comprehensive PK profile, let alone a comprehensive PK profile together with a comprehensive PD profile. In some cases, limiting the number of samples that can be collected is due to safety concerns relating to blood volume issues, and in other cases, the number of PK/PD samples is limited to avoid conflict with important and complex safety and efficacy parameters that need to be assessed in the study. Nevertheless, these studies can be an integral part of the PK/PD story that will be told during evolution to a successful new drug application. In these cases, it may be appropriate to conduct sparse sampling or to measure peak and trough concentrations at various times during the course of therapy. Often, this information, together with data from other, more intense sampling in smaller patient PK/PD studies, can form the foundation for a comprehensive PK/PD profile.

Sometimes, particularly in the cases of biotechnology-derived products, the administered drug and circulating endogenous or dietary substances are one and the same. With respect to bioequivalence assessment or true PK profiling, this would create tremendous difficulties. However, for predictive PK/PD modeling and simulation, total endogenous/exogenous drug concentrations drive PD responses. There is no need to separate naturally occurring drug from exogenously administered drug to be able to accurately model responses. In fact, although it is important to separate naturally occurring analyte from exogenous analyte to characterize and validate bioanalytical methods, from a PK/PD perspective, both endogenous and exogenous analyte contribute to the resulting response [10,11].

III. PHARMACODYNAMIC INPUT

Biomarkers and surrogate endpoints come in two general forms: biochemical/-molecular markers and clinical markers [12,13]. Clinical markers include such things as pulmonary function test results, psychomotor test results, visual analog test results, nuclear imaging, and others. Although most of the chapters in this book focus on biochemical markers, both biochemical and clinical markers are governed by the same principles and both can be used for PK/PD modeling and clinical trial simulations [12,14,15]. To be quantitative measures (Fig. 2), biochemical and clinical markers need to conform to the requirements for content and GLP-like validation.

There are many variables to consider when selecting a biochemical/molecular-marker sampling site. This is a critical decision when anticipating the future of a complete clinical drug development program. On one hand, it would be ideal to sample as close as possible to the site of the biochemical event that reflects the

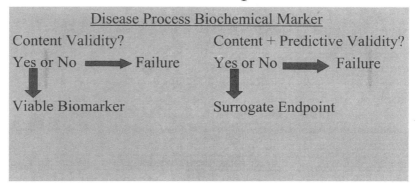

Figure 2 Quantitative use of biomarkers for PK/PD modeling requires analytical method validation whereas surrogate endpoints require method validation as well as predictive validation. Content validity ensures that the analytical method measures what it is supposed to measure. Predictive validity ensures that the surrogate endpoint predicts a later clinical endpoint.

disease state and how therapeutic intervention will impact the disease process. At the same time, this site is not likely to be readily accessible. In addition, tissue assays are not readily validated. Finally, the medical community and regulatory agencies must accept the biomarker and the appropriateness of the sampling site. Although there are a lot of variables, it is most likely that blood or urine sampling meets most acceptance criteria. If blood drug concentration assessments are acceptable for establishing bioequivalence and, therefore, therapeutic equivalence, blood biomarker concentrations should be just as acceptable for assessing effects of diseases as long as there is sufficient sensitivity to measure the biomarker in the biofluid of choice. Blood biomarker concentration-time profiles reflect biomarker concentrations at the site of disease as well as blood drug concentrations reflect drug concentrations at the site of drug action.

If blood and/or urine is selected as the best choice to evaluate the biomarker, the next question is whether the analytical methods will be sensitive and precise enough to accomplish the job. Unless the site of the disease process is the blood or urine, it is likely that biomarker concentrations will be diluted relative to those at the site of disease/drug action before reaching these sampling sites. If the method is sensitive enough, the next issue is how reproducible are the biomarker concentrations from hour to hour, day to day, month to month, and year to year during good health and in the disease state? Is there a circadian rhythm or are biomarker concentrations reasonably constant during a 24-h period following placebo, and are biomarker concentrations reproducible between two placebo control periods? How much of a change will take place in the biomarker between health and disease state and between untreated disease and disease following therapeutic intervention? Is the analytical method precise enough to quantify a meaningful change in the biomarker concentrations? Finally, will a change in the biomarker concentration actually predict a future outcome with reasonable certainty? These are questions that need to be asked and answered during the biomarker development program within the drug development process.

What changes in the biomarker concentrations are needed to be able to establish a link between the biomarker and the clinical endpoint if the biomarker is to become a surrogate endpoint? If, for example, healthy subjects exhibit biomarker concentrations in a 100–200-pg/mL range and patients exhibit biomarker concentrations in a 200–300-pg/mL range, it will probably be difficult to conclusively show the effects of therapeutic intervention even with robust bioanalytical methods. In contrast, if the concentration range is 500–800 pg/mL in disease, showing the effect of therapeutic intervention should be much easier as long as the bioanalytical method is equally robust. Both scenarios may reflect a good predictive marker, but clearly the latter example will be easier to apply. Analytical precision as well as day-to-day reproducibility in the biomarker is critical to its usefulness in drug development. This topic has been addressed in some detail in other chapters.

Biochemical PD inputs suffer from the same types of issues as PK inputs, and then some. Being able to quantify biomarkers in plasma and urine sampling has the advantage of ready access for repeat sampling without too much impact on other clinical measurements. Although the site of disease is generally not in the blood or urine, biomarker concentrations in these matrices will correlate with biomarker concentrations at the disease site just as drug concentrations in blood and urine correlate with drug concentrations at the site of action. Obviously there would be advantages for measuring the biomarker at the disease site, but multiple sampling would be difficult for cases where the disease site is not readily accessible. However, if the opportunity presents itself, it would be wise to sample at the disease site on a few occasions to correlate peripheral sample results with those from the disease site. Sparse sampling at both sites would help to create a bridge between the two sites and start bridging the biomarker in blood and/or urine to the ultimate clinical endpoint (Table 2 and Fig. 3).

Often a desirable biomarker is the same molecule that was identified in the PK input section as an endogenous ligand. For example, when angiotensin II antagonists compete with angiotensin II at the receptor site, angiotensin II becomes a principal biomarker of interest to characterize the renin–angiotensin system [5].

IV. PK/PD MODELING CONCEPTS, STRATEGIES, AND DESIGNS IN DRUG DEVELOPMENT

This chapter will not focus on the various types of response such as graded, categorical, survival, or frequency that can be encountered during drug

Table 2 Linking Biomarkers to Clinical Endpoints to Create Surrogate Endpoints

Theoretical foundations for mechanism-based disease process and impact of therapeutic
   intervention
Experimental foundations for mechanism-based disease process and effect of
   therapeutic intervention
   Preclinical
      In vitro binding to enzyme/receptor
      In vivo in animal models (transgenic)
   Clinical
      Healthy human subjects in phase I
      Healthy human disease models in phase Ib
      Human patient subjects in phase IIa
      Human patient subjects in phase IIb/III
      Epidemiological evidence
Previous clinical experience with therapeutic class
Previous clinical experience with biomarkers
Simulated biological systems

Linking Biomarkers to Clinical Endpoints

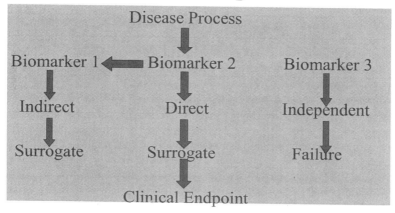

Figure 3 Linking biochemical markers to clinical endpoints creates surrogate endpoints. If biomarker 2 functions in a direct biochemical path to the clinical endpoint or biomarker 1 is a breakdown product in the cascade that leads to the clinical endpoint, it is a surrogate endpoint. Biomarker 3 is not a surrogate because it is not involved in the biochemical cascade to the outcome.

development or the types of models including direct or indirect link, direct or indirect response, hard or soft link, time-variant or time-invariant and reversible or irreversible response. These topics have been addressed in detail in an earlier publication [16]. This chapter focuses on concepts, strategies, and designs for modeling PK/PD relationships based on drug concentrations and biomarker concentrations in biological fluids [17–20].

The time to start thinking about biomarkers and their use for PK/PD modeling is when the pharmaceutical research and development division starts thinking about a new therapeutic target. Mechanism-based drug development requires an in-depth knowledge of the disease process and how therapeutic intervention can alter that process. During discovery, the receptor or enzyme that is believed to mediate the disease process is used for screening. Preclinical testing uses models including induced-disease and transgenic animals to extend the mechanism and to test the safety and tolerance in whole organisms. Early clinical development takes the theoretical rationale and preclinical experience into healthy and/or patient subjects to test for safety, tolerance, and initial proof of concept or principle. Later clinical development applies all previous experience to the final confirmatory proof of safety and efficacy studies. Each development stage can involve PK/PD modeling and simulation to improve the process by communicating and transferring knowledge within as well as between teams in the development continuum. PK/PD modeling is a tool that allows information that is gained during earlier development to be converted to knowledge that can be applied during later stages in the development process. In addition, PK/PD modeling and simulated predictions can help to improve communication with management teams that will ultimately need to act on the recommendations from the project team.

A central question during strategy development is "What are we going to do with the data?" If the data are going to be used to determine what markers may be different between disease and health, a clinical chemistry normal versus abnormal range approach may be appropriate. In contrast, if quantitative information is needed to establish concentration–effect relationships and to link biomarkers with clinical endpoints in a quantitative manner, the rigor associated with a GLP-like assay similar to a drug assay may be required.

In addition, as described earlier, underlying biomarker concentrations in health and untreated disease must be well documented to establish what changes will be needed to show a pharmacological effect. Since most PK/PD modeling is conducted for at least 24 h after a single dose or during a dosing interval during repeat dosing, baseline values in health and in disease should be determined over the intended sampling interval. Testing the biomarker profile during the intended sampling schedules will establish the potential for circadian rhythms or other trends in biomarker concentrations. In addition, placebo studies should be conducted so active treatment results can be compared with placebo control results in the same subjects or at least a group of subjects under the same circumstances.

The conceptual framework for PK/PD modeling should start at the point of target selection and continue through the entire drug development program. PK/PD modeling and simulation can be used to predict future outcomes and to communicate PK/PD concepts to stakeholders in the enterprise. The initial PK/PD modeling strategy should be planned as soon as the target has been identified. Models are simplified mathematical descriptions of complex systems that attempt to focus on the critical elements of the complex system while minimizing other variables. To create acceptable models, the model builders need to work with other researchers in the development process to (1) clearly identify the question(s) that needs to be answered, (2) explicitly state the underlying assumptions in the model, (3) explain why the specific model was selected at the exclusion of others, and (4) test the model system to ensure that it performs the way it was intended.

Once in vitro data and initial animal PK data are available, simulated dosing regimens can be tested to optimize delivery to the site of action. Is it possible to achieve the desired effect in whole animals? Even if it is possible in animals, will it be feasible in humans? To achieve the objective in humans, will it require QID dosing when the marketing group has stated that only QD dosing is acceptable? Is it time to start thinking about a controlled-release dosage form before entering into patient trials?

Single intravenous doses in healthy or patient subjects can be used to accurately control both the amount of drug reaching the systemic circulation and rate of drug input to better understand PK/PD relationships. The intrinsic PK/PD relationship can be evaluated after an intravenous bolus dose. This intrinsic relationship can be evaluated by modifying the input rate using first-order intravenous infusions [18,20,24,25]. PK and PD transfer rates can be controlled and evaluated using first-order intravenous infusions. First-order intravenous delivery is the only way to ensure that the PK input rate can be adjusted to determine when PK input rate causes a change in the PD on/off rate. As a result, the PK/PD relationship can be studied through designs that allow the investigator to isolate various PK and PD parameter values.

Although the influence of input function on PK/PD relationships can be seen in effect vs. concentration hysteresis loops and effect vs. time profiles, it is most obvious and quantifiable using concentrations in the effect compartment from direct or indirect link PK/PD models or from receptor association/dissociation rate transfer models. Following a bolus intravenous dose, the initial phase of concentrations in the effect compartment reflect the rate-controlling transfer constant to the effect site as long as the transfer rate constant is greater than the terminal elimination rate constant. If the same size dose is administered via a first-order intravenous infusion and the rate of infusion is slowed, it will begin to influence the initial phase of the effect concentration time curve as the input rate becomes slower than the rate-controlling transfer rate constant. The actual rate constants for this initial input phase can be estimated using the method of residuals. The initial rate will not change until the first-order input rate becomes slower than the transfer rate in the direct/indirect models or the composite rate of association/dissociation from the site of action.

Results from each study provide information that can be assimilated, integrated with other information, and then applied to the next study that creates better study designs and development strategies. In addition, PK/PD modeling results can be used to simulate/predict PK/PD results for the next study. As the process moves along, the predictive capability of the evolving PK/PD models should improve as the number of assumptions needed to create the model decreases. Assumptions decrease as knowledge-yielding information increases. For example, before moving into phase IIb/III, it might be prudent to simulate the influence of dose size, dose regimen, inclusion/exclusion criteria, and clinical endpoint sampling times on a study's ability to show a dose–response relationship as well as a difference between active and placebo treatment. Will one dose fit all? Do men and women require different doss? Is age a factor in the population of interest?

Another approach is to conduct population analyses across studies using all PK data, all PK/PD data, and all PD data when available and using sparse sampling as needed and optimum sampling strategies whenever possible. Including all credible data in these analyses makes for a better profile of the PK/PD properties of the drug and supplements traditional PK/PD profiles from small study groups. The composite PK/PD profile should be able to describe the entire PK/PD database.

A. Phase 0–I

Comparative in vitro human and animal absorption data plus pharmacokinetic and pharmacodynamic data from several animal species can be used to predict first human exposure results using allometric scaling [21]. In addition, in certain drug classes, the mechanism of action and proof of concept can be evaluated in healthy subjects [5,22,23].

In these cases, in vitro receptor/enzyme-binding data from both animals and humans can be correlated and then used to predict human response during the single dose, first in human studies. Once single-dose data are available, they can be assimilated, integrated with what is already known about the new molecular entity (NME), and applied to the multiple-dose study to predict safety, tolerance, and PK/PD results from various doses and regimens before the trial is initiated. If the simulated results do not look promising, the development team may want to (1) conduct the study as originally designed to confirm or negate the simulated outputs, (2) modify the dosing schedules based on the simulated outputs in an effort to improve the NME's safety and efficacy profile, or (3) conduct some combination of original design and a modified design to test the simulated results and make an effort to improve the overall profile. Money spent here can only improve later development if the NME survives phases I and IIa. If the mechanism of drug action can be tested in healthy subjects, earlier answers to several questions can be obtained. Some examples include: Is there a clinically significant food effect? Are there clinically significant differences between genders or as function of age? Is observed intrasubject variability sufficient to warrant discontinuation of development? Expanding the use of PK/PD models using biomarkers earlier in the development process is critical to successful, effective, and efficient development programs by killing unwarranted NMEs earlier while building a solid, medico scientific foundation on which to design confirmatory proof of safety and efficacy trials for more promising NMEs.

B.   Phase I–IIa

If the phase I PK/PD program answers all of the questions that it is intended to answer, the transition from phase I to IIa simply needs to address the impact of the targeted disease on PK and PD as well as measure biochemical markers and/or clinical endpoints to determine whether therapeutic intervention actually works to alter the disease process. Phase IIa results together with the PK/PD models that describe them become the foundation for predicting what will happen in expanded populations and alternative disease states. At the same time, this is the most critical juncture for go/no-go decisions, kill or be killed. Although each go/no-go decision is important for the potential success or failure of a pharmaceutical company, the end of phase IIa is the last chance to make the correct decision before moving into the huge investment arena. If a questionable NME is moved beyond phase IIa, resources will be allocated to the wrong development program. For drug development to be both efficient and effective, it is critical to discontinue development of NMEs that will have questionable likelihood of success in development or in the marketplace. PK/PD modeling and simulation using appropriate biomarkers, surrogate endpoints, and clinical endpoints are the tools that allow pharmaceutical companies to make correct decisions and ultimately to deliver new therapies that will fuel future research.

Study designs that can be used to sort out variables that can impact PK/PD study results include: (1) a crossover wherein one group of subjects receives placebo followed by active treatment whereas the other group receives active followed by placebo; (2) an embedded crossover wherein all subjects receive placebo in period 1, half of the subjects receive active in period 2 and placebo in period 3, and half receive placebo in period 2 and active in period 3; and (3) a cascade design wherein all subjects receive placebo treatment in period 1, active treatment in period 2, and placebo again in period 3. Obviously these designs can be modified to alter the amount and type of information that can be obtained. There are limitations to each of these designs, but they do yield placebo and active data for all subjects. In each design recovery/carryover data can be obtained. Embedded crossover designs are useful in situations where stress associated with confinement and procedures may alter biomarker concentrations [26].

C.   Phase IIa–IIb/III

Once all of the phase 0, I and IIa data have been collected, evaluated, assimilated, and integrated using PK/PD modeling to create the comprehensive PK/PD profile, this knowledge can then be applied to the design of the phase IIb/III program. Phase IIb/III is where the largest investment takes place. Select the promising compounds for continued development while killing those compounds that do not show promise and the company will live to develop more compounds in the future. If the process is flawed and compound selection is poor because inadequate and/or inappropriate biomarkers and PK/PD study designs were used to create an unstable foundation, phase IIb/III development programs will be saturated with compounds, only a few of which will succeed. Therefore, the cost of doing business skyrockets.

If the disease-process mechanism, drug-action mechanism, and concentration–effect relationships are all understood by the end of phase IIa, clinical trial simulations should provide tremendous insight into potential phase IIb/III study designs. To ensure that the phase III program can be optimized, phase I and IIa studies must be designed to answer questions about these mechanisms and relationships and to replace assumptions in earlier models with PK, PD, and PK/PD answers before moving into phase III.

Poor correlation between a biomarker and a clinical endpoint is almost always attributed to the biomarker. Why is poor correlation the fault of the biomarker? Except for clinical endpoints, such as a cure as opposed to continued disease or life as opposed to death, what makes clinical endpoints flawless? The answer is nothing. Clinical endpoint measurements are often extremely variable and require large numbers of observations to establish a trend. In some cases, the selected clinical endpoints may not adequately address the disease at hand. Perhaps the lack of agreement between certain biomarkers and certain clinical endpoints is the fault of the biomarker or the lack of agreement between the biomarker and clinical endpoint is caused by poor process validation for the measurement of the biomarker, the clinical endpoint, or both. However, the lack of agreement between the biomarker and clinical endpoint may also be caused by selection of an unreliable clinical endpoint or an unreliable clinical endpoint measurement. When searching for a surrogate endpoint, the clinical endpoint must be questioned and tested as rigorously as the biomarker (see Table 2). Many clinical endpoints are founded in history rather than science. History and science are not one and the same.

V. PK/PD MODELING AND SIMULATION IN REGULATORY DECISION MAKING

This chapter is not about regulatory decision making from a regulatory perspective. Three preceding chapters have already provided some insight into the regulatory perspective. This chapter is about regulatory decisions from one drug developer's perspective. Regulations are in place, so let's use them. It is time to buck the conservative old-school trend, to continue doing the same old things the same old way because they have worked in the past. You can bet that the old ways are not going to continue to work in the future. The competition is not going to get any easier. If you want to be a player in the new game, you will need to adopt a new philosophy.

A PK/PD model is a simplified quantitative mathematical relationship between exposure and response(s). PK/PD models in the regulatory environment should facilitate market access and labeling (see Table 3). Models should support bridging one population to another like adults to pediatrics and approval in one region to approval in another or supporting a switch from one dose/regimen to another. Biomarkers and PK/PD modeling should be used to support accelerated approval via fast-track provisions as stated in 21 CFR, Section 112. Strong PK/PD modeling support should also warrant approval based on a single pivotal proof of safety and efficacy as stated in 21 CFR, Section 115. PK/PD modeling is a tool that needs to be more widely applied to improve therapeutics through a variety of approaches within the pharmaceutical industry, the agencies that regulate it, and the health care professionals who use their products, as shown in Fig. 4.

VI. PK/PD MODELING AND SIMULATION IN THE FUTURE

No living organisms will be exposed to new therapeutic agents until they are approved for worldwide marketing because PK/PD modeling and simulation will replace human testing. Humans will be genetically altered so drugs will behave exactly the same way in each and every individual. Drugs will not be used to treat

Table 3 Using Biomarkers and Surrogate Endpoints

Early internal decision making
Early decision making with regulators
Generic approvals
Accelerated approvals
    Cancer
    HIV/AIDS
    Others
Regular approvals
    Diabetes
    Hypercholesterolemia
    Hypertension
    Osteoporosis
    Ulcers
    Cancer
    Others

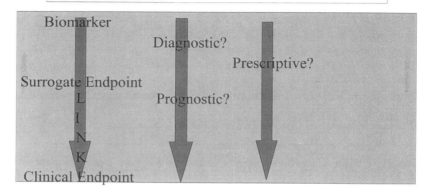

Figure 4 Biomarkers can evolve to serve several functions. As described earlier, biomarkers can be surrogate endpoints if they are linked to the clinical endpoint. In parallel, they can become diagnostic agents if they are used to diagnose disease and prognostic agents if they are used to evaluate disease progress and the effect of therapeutic intervention. In select cases, the biomarker can be used for diagnosis and to determine appropriate therapeutic intervention.

disease because there will be no disease. Rather, drugs will be used to make people look like they are 18 years old until their 250-year birthday. These drugs will all be natural, without potential adverse effects, much like herbal supplements are promoted today.

PK/PD modeling and simulation will play a critical role in the future. Drug targets, drug design, drug testing, and labeling will be done using modeling and simulation technology (in silico). Once the new drug is marketed, large postmarketing safety and efficacy trials will be conducted to ensure that the in silico predictions were, in fact, correct. And, of course, they will be. Safety and efficacy information will be captured from an implanted biosensor that monitors body functions and biochemical markers and downloads the results to a central Food and Drug Administration computer. This is the future of PK/PD modeling and simulation . . . but not in my lifetime.

In the interim, we are getting a lot of new information from genomics, functional genomics, proteomics, and functional proteomics. Even with genomics, functional genomics, and proteomics, our understanding of disease processes is critically deficient. We may know where to intervene, but we still do not know how that intervention will play out later in that cascade or in another resultant cascade, or in both. Adverse events that are extensions of beneficial effects or totally separate but parallel processes need to be incorporated into our knowledge base. Once we understand the disease process, the mechanism for intervention, and the mechanism for adverse events, how will we build that knowledge into our models, which are simplistic by definition? Clearly, the models must become more sophisticated and comprehensive, but hopefully not much more complex. If we make the models too complex, they will not be useful for communicating with stakeholders and, therefore, will not be believable. Clearly, simplified disease/health-state models are needed. Now, we need expanded, more effective bioinformatics systems to help us pull all of the information together so we can assimilate, integrate, and apply it to the drug development process through simplified models.

Once these objectives have been accomplished, our next objective is to take information from earlier phases of drug development and convert it to knowledge that allows us to design, conduct, and evaluate the next phase of drug development more efficiently. In many cases, the missing critical element is the ability to communicate from one area of expertise to another. PK/PD modeling and clinical trial simulation can be that communication tool. Appropriate use of PK/PD modeling and clinical trial simulation can improve drug development and, therefore, therapeutics. In my lifetime, we may see extrapolation of in vitro cell culture experiments to minimize the need for certain preclinical and clinical studies (see Table 4).

Starting today, we need to be able to assimilate and integrate preclinical and clinical information so it can be used to reduce the number of assumptions in our models and thereby improve our predictions. To accomplish this, we need to develop better understanding of both the mechanistic basis of the disease process and the mechanistic basis for therapeutic intervention at the molecular and cellular level. Today, most of the models used to predict phase III study results are based more on assumptions than on reliable knowledge based on experimental results. We need to populate our phase III models with real parameter values and distributions from clinical studies [27,28]. Although there is a long way to go, progress is being made and the tools are becoming available to make it happen.

PK/PD models and clinical trial simulations are tools for drug development. The missing element in many companies today is the team of researchers who can provide appropriate inputs and guide output selection to create drug development knowledge. The PK/PD modeling team will need to

Table 4 Do Not Confuse Information with Knowledge

Information is power . . . No, KNOWLEDGE is power.
Assimilate, integrate, and successfully apply information to create knowledge.

routinely work with synthetic chemists, pharmacologists, toxicologists, bioanalytical chemists, biostatisticians, clinical pharmacologists, regulatory affairs specialists, and therapeutic area experts. And, in specific cases, formulation scientists may need to be brought to bear to achieve the PK/PD objectives of drug delivery. This team will be responsible for the modeling and simulation as well as effectively communicating the results to the remainder of the organization. In many cases, pharmaceutical companies are not structured to accomplish this objective because the right resources are not available and/or the need has not yet been identified at the top.

REFERENCES

1. Emery, P.; Luqmani, R. The validity of surrogate markers in rheumatic disease. Br. J. Rheumatol. **1993**, *32* (S3), 3–8.
2. Prentice, R.L. Surrogate endpoints in clinical trials: definition and operational criteria. Stat. Med. **1989**, *8*, 431–440.
3. Rolan, P. The contribution of clinical pharmacology surrogates and models to drug development—a critical appraisal. Br. J. Clin. Pharmacol. **1997**, *44*, 219–225.
4. Colburn, W.A. Surrogate markers and clinical pharmacology. J. Clin. Pharmacol. **1995**, *35*, 441–442.
5. Colburn, W.A. Optimizing the use of biomarkers, surrogate endpoints and clinical endpoints for more efficient drug development. J. Clin. Pharmacol. **2000**, *40*, 1419–1427.
6. Bennett, D.A.; Waters, M.D. Applying biomarker research. Environ. Health Perspect. **2000**, *108*, 907–910.
7. Atkinson, A., Jr.; Colburn, W.A.; De Gruttola, V.; et al. (Biomarkers Definitions Working Group). Biomarkers and surrogate endpoints: preferred definitions and conceptual framework. Clin. Pharmacol. Ther. **2001**, *69*, 89–95.
8. De Visser, S.J.; Van Der Post, J.; Pieters, M.S.M.; Cohen, A.F.; Van Gerven, J.M.A. Biomarkers for the effects of antipsychotic drugs. Br. J. Clin. Pharmacol. **2001**, *51*, 119–132.
9. Eldon, M.A. Clinical pharmacokinetics during drug development: strategies and study designs. Appl. Clin. Trials **1996**, *5* (10), 56–64.
10. Colburn, W.A.; Gibson, D.M. Pharmacokinetic/pharmacodynamic modeling of baseline effects: influence of endogenous agonists. In *Pharmacodynamic Research— Current Problems and Potential Solutions*; Kroboth, P.D., Smith, R.B., Juhl, R.P., Eds.; Harvey Whitney Books: Cincinnati, OH, 1988; 167–184.
11. Colburn, W.A. Drugs and endogenous ligands compete for receptor occupancy. J. Clin. Pharmacol. **1994**, *34*, 1148–1152.
12. Colburn, W.A. Selecting and validating biologic markers for drug development. J. Clin. Pharmacol. **1997**, *37*, 355–362.
13. Blue, J.W.; Colburn, W.A. Commentary: efficacy measures: surrogates or clinical outcomes? J. Clin. Pharmacol. **1996**, *36*, 767–770.

14. Sim, J.; Arnell, P. Measurement validity in physical therapy research. Phys. Ther. **1993**, *73*, 102–115.

15. Lee, J.W.; Hulse, J.D.; Colburn, W.A. Surrogate biochemical markers. Precise measurement for strategic drugs and biologics development. J. Clin. Pharmacol. **1995**, *35*, 464–470.

16. Derendorf, H.; Lesko, L.J.; Chaikin, P.; et al. Pharmacokinetic/pharmacodynamic modeling in drug research and development. J. Clin. Pharmacol. **2000**, *40*, 1399–1418.

17. Colburn, W.A. Pharmacokinetic/pharmacodynamic modeling: study design considerations. In *Pharmacokinetics and Pharmacodynamics: Research, Design and Analysis*; Smith, R.B., Kroboth, P.D., Juhl, R.P., Eds.; Harvey Whitney Books: Cincinnati, OH, 1986; 65–84.

18. Colburn, W.A.; Eldon, M.A. Models of drug action: experimental design issues. In *The In Vivo Study of Drug Action*; Van Boxtel, C.J., Holford, N.H.G., Danhof, M., Eds.; Elsevier Science Publishers: New York, 1992; 17–29.

19. Colburn, W.A.; Eldon, M.A. Simultaneous pharmacokinetic/pharmacodynamic modeling. In *Pharmacodynamics and Drug Development: Perspectives in Clinical Pharmacology*; Cutler, N.R., Sramek, J.J., Narang, P.K., Eds.; Wiley: London, 1994; 19–44.

20. Colburn, W.A. Decision making during new molecular entity development. Appl. Clin. Trials **1996**, *5* (10), 44–55.

21. Obach, R.S.; Baxter, J.C.; Liston, T.E.; Silber, B.M.; Jones, B.C.; MacIntyre, F.; et al. The prediction of human pharmacokinetic data from preclinical and in vitro metabolism data. J. Pharmacol. Exp. Ther. **1997**, *283*, 46–58.

22. Heath, E.C.; Pierce, C.H. Inducing disease in healthy volunteers for early evaluation. Appl. Clin. Trials **1999**, *8* (5), 42–48.

23. Carlson, A.; Hulse, J.D.; Faulkner, R.D.; Johnson, F.K.; Pederson, J.; Lee, J.W. Application of clinical assays for pharmacodynamic measurement of coagulation and fibrinolysis markers. Pharm. Res. **1997**, *14*, S518.

24. Colburn, W.A.; Brazzell, R.K.; Holazo, A.A. Verapamil pharmacodynamics after intravenous and oral dosing: theoretic consideration. J. Clin. Pharmacol. **1986**, *26*, 71–73.

25. Colburn, W.A. Commentary: toxicology evaluation and single intravenous dose studies in human subjects. J. Clin. Pharmacol. **1996**, *37*, 4–6.

26. Colburn, W.A.; Gottlieb, A.B.; Koda, J.; Kolterman, O.G. Pharmacokinetics and pharmacodynamics following intravenous bolus and infusion doses of AC137 (25,28,29 tripro-amylin, human) to insulin dependent diabetics. J. Clin. Pharmacol. **1996**, *36*, 13–24.

27. Kimko, H.C.; Reele, S.S.B.; Holford, N.H.G.; Peck, C.C. Prediction of the outcome of a phase 3 clinical trial of an antischizophrenic agent (quetiapine fumarate) by simulation with a population pharmacokinetic and pharmacodynamic model. Clin. Pharmacol. Ther. **2000**, *68*, 568–577.

28. Veyat-Follet, C.; Bruno, R.; Olivares, R.; Rhodes, G.R.; Chaikin, P. Clinical trial simulation of docetaxel in patients with cancer as a tool for dosage optimization. Clin. Pharmacol. Ther. **2000**, *68*, 677–687.

9

Pharmacogenomic Biomarkers

Richard D. Hockett and Sandra C. Kirkwood
Eli Lilly and Company, Indianapolis, Indiana, U.S.A.

I. INTRODUCTION: GENETICS AS BIOMARKERS

Recently, two independent groups released the first draft of the human genome [1,2]. This astounding achievement has resulted in a slate of editorial comments about how this knowledge will revolutionize medicine and drug development. Identification of genes that cause or modify risk for disease and those that affect response to therapy or predict the development of a side effect have been espoused as a means to improve therapeutic outcome [3]. The term "the right drug into the right patient" has often been used to describe the effects the genetic revolution will have on drug development. While studying the human genetic code undoubtedly will reveal many secrets that ultimately will impact drug discovery and development, maturation of genetic associations into successful genetic biomarkers requires multiple, time-consuming steps. We are at the beginning of a long pursuit whose ultimate goal is improved patient care. However, for most disease states, genetic biomarkers to identify patients at risk for that disease, stratify patients by clinical outcome, indicate treatment response, or predict adverse event occurrences are in reality, several years away.

The development of any successful biomarker requires the completion of several steps including laboratory and clinical studies (Fig. 1, steps A, B, C). The end result of the process is a biomarker that will stratify patients by predisposition to disease, by likely response to therapy, or by susceptibility to an adverse event. The marker must be sufficiently validated with the risk conferred by the marker sufficiently understood. In addition, a validated assay, with defined sensitivity and specificity, must be generally available to meet the standards necessary for modern medicine. If the biomarker is needed to aid in the development of a drug

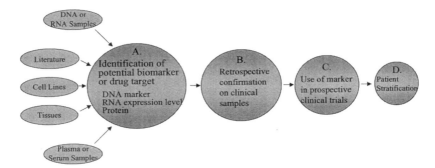

Figure 1 Schematic development of a genetic biomarker. (A) A biomarker is first identified as associated with the phenotype in the laboratory. Several sources for this link are identified above. (B) Once the genetic marker has been associated with the phenotype in the laboratory, the association is confirmed in samples from patients who have the phenotype or trait of interest. This may be stored samples or samples from a clinical trial not prospectively stratified by the biomarker. (C) If part B is successful, then the marker is used prospectively in additional clinical trials to validate its use. (D) The ultimate goal of this approach is to stratify patients by the biomarker, into groups by response to drug or susceptibility to side effect development.

or is to be utilized to drive prescribing practices, then regulatory approval will be mandatory.

To navigate this tortuous route to approval and acceptance, the marker must first be identified and validated in the research laboratory (Fig. 1, step A). The marker's underlying science and relationship with phenotype must be thoroughly tested, reproduced, and validated. A variety of methodologies are utilized in this phase often culminating in an animal model to mimic human disease. The goal of this phase is to properly frame the hypothesis, so that analysis of clinical specimens can adroitly associate the marker with the desired phenotype.

The next step is the retrospective analysis of the biomarker in a cohort of stored clinical samples (Fig. 1, step B). Alternatively, clinical association studies can be performed to evaluate the association of the biomarker with the outcome or phenotype but where the marker is not used for the clinical management of the patient. Both methods will be referred to herein as retrospective analyses. The ultimate goal of these retrospective studies is a definitive association of the marker with a phenotype, drug effect, or disease susceptibility. This association provides the initial clinical evidence confirming the previous scientific research and validating the hypothesis. Clinical association studies also justify the next step, prospective clinical trials utilizing the biomarker in the evaluation or clinical management of the patient (Fig. 1, step C).

The failure of the validation of many biomarkers at this stage is often because of a poorly defined hypothesis, weak scientific evidence that inadequately links the marker to the phenotype, or statistical issues such as a deficiency in study design and power. Furthermore, depending on the strength of the scientific evidence supporting the association and risk attributable to the marker, multiple definitive clinical association studies may be required before investigators will be willing to use the marker in prospective patient management. Another consideration prior to utilizing the marker for patient management is understanding the general applicability of the marker in a variety of populations stratified by ethnicity.

The final step is the prospective stratification of patients into phenotypic groups based on the biomarker (Fig. 1, step D). Although this chapter will discuss genetic biomarkers of disease susceptibility, the primary focus of the chapter is on the utilization of genetics to develop rational therapeutics. For clinical drug development, the stratification is usually into groups likely to respond to a particular therapy, or to experience an undesired reaction to a drug. In this sense, genetics is no different than any other biomarker: it is simply a different kind of blood test.

For purpose of this chapter, genomic biomarkers will be stratified into two main categories, disease susceptibility and pharmacogenomic (Table 1). Genetic markers for disease susceptibility have a large impact in medicine, with many rare disorders explained by single gene mutations (see below). However, to date few of these genetic associations have led to the introduction of new therapeutics. The difficulty lies in relating the specific gene mutation to a "druggable" target. The identification of genes influencing complex traits such as cancer, heart disease, and drug response has great potential to provide future drug targets and biomarkers. The most widespread application of genetic markers in drug development and medicine today is the genotyping of the drug metabolism enzymes of the liver. These and the other examples outlined below illustrate the infancy of this field, but highlight the potential of pharmacogenomics.

II. DISEASE SUSCEPTIBILITY

A. Single-Disease Genes with Mendelian Inheritance

For years identification of germline single-gene mutations predisposing to rare diseases has been possible. The vast majority of these mutations are inherited in a mendelian fashion, dominant or recessive, with high penetrance, often resulting in a definitive disease phenotype. Until the advent of modern genetic techniques, the primary biomarkers were assays detecting the abnormal protein product of the mutated gene. Examples of this include testing for the single nucleotide change leading to sickle cell anemia as performed by the sickledex test [4] or testing for

Table 1 Types of Applied Genomic Biomarkers

I. Disease susceptibility genes
     1) Single-disease genes (mendelian inheritance)
        a) Huntington's disease
        b) Cystic fibrosis
        c) Duchenne muscular dystrophy
        d) Factor V Leiden
        e) CCR5 chemokine
     2) Genetic associations in complex diseases
        a) Coronary artery disease
        b) Diabetes
        c) Osteoporosis
        d) Mental illness such as depression
        e) Neurological diseases such as Parkinson's and Alzheimer's diseases
     3) Genetic changes associated with tumorgenesis
        a) Inherited mutations
           BRCA1 and *BRCA2* mutations
           FAP/HNPCC
           GSTM1 mutations in bladder cancer
        b) Spontaneous mutations in the tumor
           p53 mutations
           c-src mutations
        c) Multiple transcript changes leading to reclassification
II. Pharmacogenomic biomarkers
     1) Genetic polymorphisms predicting drug metabolism
        a) Cytochrome P450 genes
        b) *N*-Acetyltransferases
        c) Glutathione-*S* transferases
     2) DNA mutations predicting drug response
        a) Herceptin in breast cancer
        b) 5-LOX and Zileutin
        c) CEBP and statins
        d) Baseline TS and response to 5-FU
     3) DNA mutations predicting adverse events

the polymorphisms associated with phenylketonuria by measuring the blood phenylalanine levels [5]. With the advent of modern molecular biology, disease genes are now identified using tests for the specific genes themselves. Examples of these include the identification of the trinucleotide repeat (CAG) expansion in the Huntington's disease gene [6] and the mutations for cystic fibrosis, most frequently the delta F508 variant [7,8]. Even though the penetrance of these disease genes is often high and the presence of the disease gene confers a high likelihood of manifesting the disease, the scientific validation of the identified mutations as the causative alleles can be quite laborious. However, once confirmed, the clinical validation of these biomarkers is relatively straightforward, applying the same principles as any diagnostic test. Specific tests for single-gene disorders are widely accessible with hundreds commercially available [9]. These tests enable individuals to know their risk for developing a disease, and may be performed following or prior to the onset of symptoms or even in utero. Therefore, genetic counseling aimed at understanding the benefits and risks should accompany such testing.

Although many of the diagnostic tests for single-gene disorders are commercially available, the majority of these single-gene diseases are rare, limiting the widespread use of these tests. Even though the sensitivity and specificity of the testing methodology is high, owing to the low prevalence of most of the associated diseases, widespread testing for these genetic markers must be carefully considered. To illustrate this point, if 10,000 individuals are tested with a particular genetic test whose sensitivity is 99% for a disease gene carried by 1 out of 10,000 people in the general population (1% chance of yielding a false-positive rate, considered a very good clinical test), 99 false-positives will be found for every 1 true-positive. Therefore, these tests are targeted only at those individuals with an elevated risk for the particular condition. Cystic fibrosis is the exception, with the carrier rate in the Caucasian population estimated at 1 in 20–25 [7,8]. The false-positive rate in this circumstance is only 1 for every 4–5 true-positives versus the 99 false-positives out of 100 total positives in the previous example for a rare disease. Therefore, the cystic fibrosis test is being considered for more widespread utility [10]. While the identification of disease-causing genes may illuminate the underlying biology and mechanism of action of the disease identifying new potential targets, none to date have yielded curative therapeutics.

B. Genetic Associations in Complex Diseases

The genetic biomarkers with the greatest potential to impact medicine are those associated with complex diseases. Complex human disorders, caused by multiple genetic and environment factors, are characterized by high population prevalence, lack of clear mendelian patterns of transmission, etiological and phenotype heterogeneity, and a continuum between disease and nondisease states [11,12]. Complex diseases cause significant morbidity and mortality, adding billions to the health care budget each year. Hear disease, diabetes, osteoporosis, depression, and Alzheimer's disease are a few examples. Identification of genetic markers for complex disease will not only assist in predicting those individuals who are predisposed to disease, but potentially have significant impact in the pharmaceutical industry by providing new targets for therapeutics. For example, the current slate of pharmaceutical products is aimed at an estimated 500 proteins. The current estimate of 30,000–60,000 genes in the human genome provides 60–120 times that number of potential targets. Either directly, or indirectly through protein–protein interactions or as part of feedback loops, the novel genes associated with multigenic diseases represent a significant pool of potential new targets.

Nonetheless, the use of genetic polymorphisms in the drug discovery process will be a long road progressing through multiple phases, similar to that outlined in Fig. 1. First, and foremost, the polymorphism must be associated with the disease. The process is complicated by the influence of multiple genetic and environmental cofactors leading to the disease phenotype, and a lack of scientific understanding of the role these factors play in the pathophysiology of disease. When the pathophysiology is unclear and candidate genes are unknown, whole genome scans to identify chromosomal regions linked to a particular phenotype are applied [13–15]. After identification of linked chromosomal regions, fine mapping to narrow the region, and association of candidate genes or genetic variants in this region with a particular disease, are necessary [13]. To verify that the identified changes in the study population have applicability to the population at large, population-based studies are conducted. Alternatives to identifying chromosomal regions linked to a particular phenotype are genomic approaches capitalizing on the technological advances in molecular biology and information systems to develop a candidate gene list. These candidate genes are then tested for association with the disease phenotype [16].

A recent example of success identifying genetic variants associated with a complex disease is the linkage and association of *CAPN10*, the gene encoding the cysteine protease calpain-10, with type II diabetes. The work by Horikawa and colleagues [17] is a significant achievement and highlights many of the process principles discussed above The inheritance of specific *CAPN10* haplotypes defined by three single nucleotide polymorphisms (SNPs) was associated with a threefold increased risk for type II diabetes in Mexican-Americans from a single country in Texas. This association of *CAPN10* was replicated in two northern European populations: a population from the Botnia region of Finland and the German population of Saxony. This work identifying calpain-10 was an extension of previous work linking regions of chromosome 2 to type II diabetes. The time to identify these associations took somewhat over 5 years, depending on how the boundaries are defined.

Evidence for a functional role of *CAPN10* in the development of diabetes includes the association of *SNP-43* in the presence of *SNP-44* with mRNA levels in skeletal muscle [17,18] and *SNP-43*'s association with measures of insulin action.[18] However, the mechanism of action of the genetic polymorphism and the role of *CAPN10* in type II diabetes remains elusive. How a ubiquitous serine protease affects glucose control and the effect of the polymorphism in the calpain-10 locus on the protein product or how this might relate to the development of type II diabetes in unclear (too much calpain-10 or too little). This uncertainty regarding the role of calpain-10 illustrates one of the complicating factors in developing biomarkers for complex traits. Once a candidate gene is identified, associated with a phenotype, and replication studies are completed, functional studies related to the disease process or development of the phenotype are necessary. These functional studies must show that the associated genetic variant results in a functional change related to the phenotype, not simply an association secondary to the polymorphism being in disequilibrium with the causative gene.

After identification of the relationship of the biomarker to the phenotype, all of the polymorphisms in that gene must be identified and their relationship to the observed functional effect understood. This analysis should include associations in a variety of populations, because a biomarker may have utility in a specific population but not have applicability to another. This is currently an active area of investigation for *CAPN10*. As an example, the results of *CAPN10* in an alternate population by Evans et al. [19] illustrate the importance of multipopulation-based confirmations. In this report, the authors were unable to replicate the association of the specific *CAPN10* alleles previously identified but rather associated additional alleles at the locus with increased type II diabetes risk in a population of Caucasians of British/Irish ancestry. The differing polymorphisms and haplotypes associated between populations may be due to multiple susceptibility alleles at *CAPN10* or different patterns of linkage disequilibrium between a common causal variant [19]. This illustrates the necessity of large population-based studies designed to provide a better understanding of the contribution of *CAPN10* polymorphisms to type II diabetes risk.

Finally, after an understanding of gene function, its relation to disease process, and applicability across populations is discovered, the gene can be investigated as a new therapeutic target. The possibilities and methodological issues are too involved to fully discuss in this chapter, and furthermore, none of these steps are trivial. Suffice it to say this is indeed a long process that will not happen in a short period of time without substantial good fortune by the investigators. As a consequence, the discovery of calpain-10 will not affect the development of new therapeutics until the relationship of this protease to diabetes is understood. The development of viable pharmaceuticals will come after this understanding, and therefore is likely 5–10 years in the future.

As the last point of discussion of genetic associations in complex disease, the utility of a genetic biomarker is influenced not only by an understanding of the pathophysiology and its relationship to the phenotype, but also by the strength of the association with the phenotype and the specificity of the association. In considering the strength of the association both the relative risk associated with the gene and the attributable risk fraction are relevant. Two examples to illustrate this point are: (1) the association of peroxisome proliferator-activated receptor-γ (PPARγ) with type II diabetes mellitus; and (2) the association of apolipoprotein E (*ApoE4*) and the presenilins (1 and 2) with Alzheimer's disease.

In the case of PPARγ and diabetes, a successful treatment modality for diabetes was discovered prior to understanding the contribution of the target to the disease. After the development of the thiazolidinediones, heterozygous mutations in the ligand-binding domain of PPARγ were identified in three individuals with severe insulin resistance [20]. These germline loss-of-function mutations in PPARγ provided compelling evidence that this receptor is important in insulin resistance and thus glucose homeostasis. The association of the common Pro12Ala polymorphism of PPARγ with type II diabetes provided further evidence for the role of this receptor in diabetes [21]. The attributable risk fraction for this specific PPARγ polymorphism is quite large as it is common in the Caucasian population and influences the development of approximately 25% of type II diabetes. However, the relative risk conferred by the allele to an individual is modest. Owing to the modest risk increase for those carrying this genetic variant; a specific biomarker for this polymorphism has little clinical utility.

A well-known example to further illustrate this point is the risk conferred by the known genetic risk factors for Alzheimer's disease. Presenilin 1 and presenilin 2 are examples of genetic markers with high relative risk but low attributable risk fraction [22]. These mutations are extremely rare with very few individuals harboring the mutation. However, for those individuals carrying the mutation the risk for developing Alzheimer's disease is very high. On the other hand, carrying the *ApoE4* allele [23] confers modest relative risk but greater significance in terms of the proportion of the population affected by this mutation. The clinical utility of a biomarker is determined not only by the magnitude of the effect of genetic biomarkers in individual patients, but in the general population. Thus, for a biomarker to have the most clinical utility, is should be characterized by both high relative risk and high attributed risk fraction.

Inherent to successful pursuit of complex disease genes is the study design including power, methodological considerations, and phenotype definition. If the phenotyping is inaccurate, the data regarding the association of a genetic variant with that phenotype are meaningless. A major challenge facing the application of genetic biomarkers is the etiological or phenotypic heterogeneity. Heterogeneity influences not only the ability to discover a biomarker but also the clinical utility of a biomarker once identified. A biomarker that is specifically associated with the phenotype of interest has more clinical utility than one associated with a range of phenotype [12]. To aid in the discovery of applicable genetic biomarkers, genetic epidemiology principles are being applied to refining phenotypic definitions and identifying the role of genetic and environmental risk factors in disease and drug response.

C. Genetic Changes Associated with Tumorogenesis

Cancer, which results from the interaction of genetic changes and environmental factors, is by definition a genetic disease. As such, studying the genetic background of the tumor with comparison to the genetic background of the patient is a powerful investigative tool. Utilizing this paradigm, several groups have defined a progression of genetic changes required for cancer formation, the multihit theory of tumorogenesis [24]. The likelihood of cancer formation for some types of cancer is significantly increased by the inheritance of the mutant form of certain genes. Examples of cancers that are clearly linked to the inheritance of mutant genes include the *BRCA1* and *BRCA2* mutations and breast cancer [25] and familial adenomatous polyposis coli gene (*APC*) mutations and colon cancer [26,27]. Inherited mutations such as these are relatively rare in the population at large; therefore, the population attributable risk conferred by these mutant genes is relative low. The impact on medicine has been an improved understanding of the mechanisms of tumorogenesis and at the individual level has led to better screening, counseling, and early detection of tumors in individuals possessing cancer-predisposing mutations. Unfortunately, knowledge of the genetics of tumor formation has not led to a multitude of new drugs to combat this progression, usually because the underlying pathophysiology and contribution to tumorogenesis of the affected gene are not fully understood.

In the general population, tumor formation occurs most frequently from mutations arising spontaneously. The most completely described progression scheme, that of colon cancer, outlines the multiple genetic mutations at various stages of polyp formation leading to adenocarcinoma [28,29]. These genetic changes include deletions, translocations, or point mutations leading to oncogene activation, tumor suppressor gene inactivation, mismatch repair inactivation, microsatellite instability, and methylation changes. Most of these mutations lead to transcriptional changes and loss of cell growth control. For example, the oncogene c-*src* associated with colon cancer has mutations leading to constitutional activation of the enzyme. Through a series of complex intracellular interactions, activation of c-*src* leads to activation of cyclin D1 and uncontrolled cell growth [30,31].

1. Targets for Drug Therapy

Some of the genes involved in tumorogeneis may be targets for drug therapy, as with c-*src*, where specific antagonists to kinase activity could potentially be employed. Although several small molecules aimed at inhibiting protein kinases implicated in tumor formation are in development, as of this writing, most genetic changes in the cascade of tumor formation are not utilized as specific therapeutic targets. The two notable exceptions are Herceptin and Gleevec. The development of the trastuzumab antibody (Herceptin, Genentech) is the most complete example of success using genomic biomarkers in general practice and will be discussed below. The other example of a therapeutic agent specifically designed at a genetic target is Gleevec for chronic myelogenous leukemia (CML). CML is characterized by a reciprocal translocation between chromosome 9 and 22 resulting in a *bcr-abl* fusion gene whose protein product demonstrates increased tyrosine kinase activity. *Bcr-abl* is a particularly attractive target as it is sufficient to cause disease and is present in over 95% of patients with CML [32]. In addition, current knowledge indicates that early in disease CML is not characterized by a multitude of other molecular abnormalities. Gleevec is a tyrosine kinase inhibitor designed to block the ability of *bcr-abl* to phosphorylate its unknown substrate [33]. The development of this compound illustrates the time-consuming process required for the development of directed therapeutic, aimed at a genetic target [34]. The process included identification of an appropriate target, development of an inhibitor to the enzyme, and identification of a lead compound by random screening. This compound was then tested in a number of preclinical models and human ex vivo studies conducted, which ultimately culminated in Phase I, II, and III clinical trials and Food and Drug Administration (FDA) approval.

Although these successful examples can be cited, the development of directed therapeutics encounters many difficulties possibly related to the problems of activation or inhibition of ubiquitous cellular process for cell growth, leading to profound toxicity. The use of mutational markers to develop rational therapeutics is in its infancy with many relatively early in development or as yet undiscovered. It will likely take many years before they are available for widespread use. Nonetheless, identified genetic mutations provide good targets and likely will continue to be exploited.

2. Tumor Classification

A potentially more timely clinical utility of gene expression changes as a consequence of genetic alteration will be in the reclassification of tumors. Current methods of classification rely upon morphology, tumor size, and, at times, cell surface protein expression. The combination of these parameters is widely used to help determine prognosis, but they have not been helpful in identifying responders or nonresponders to various therapy modalities. The hypothesis and rationale for this line of investigation is that a more comprehensive assessment of expressed genes would improve tumor classification. Inherent in this justification is that better classification of tumors will identify those patients likely to respond to particular treatments. Or conversely, and perhaps more beneficial, suggest additional targets for new therapeutics.

The most inclusive technique utilized for gene expression analysis, array profiling, monitors multiple genes simultaneously. As a result of the sequencing of the human genome and recent technological advances including the advent of the gene chip, we are now able to monitor several thousand genes simultaneously. A gene chip is similar to a microchip but it is coated with DNA rather than electronic circuitry. Thousands of strands of reference DNA are synthesized on the chip with photolithography, ink-jet spray, or pin spotting [35]. RNA from the tissue or cell source is utilized to synthesize cDNA, the cDNA transcripts are labeled, and the mixture hybridized onto the chip. Since the RNA being tested is not selected, this analysis of total RNA has been termed the transcriptome, and the profile has been termed a "molecular fingerprint." Recent reports have used array profiling in the classification of B-cell lymphomas [36,37], epithelial ovarian cancer [38], and breast cancer [39]. In these reports, total RNA from each tumor was hybridized to gene chips with thousands of human sequences and the tumors classified based on the expression profile.

The drawback for array profiling, as with any novel technique, is the newness of the technology. For these techniques to have utility in a clinical setting, significantly more data are needed to understand their capabilities and limitations, and evaluate their reproducibility. While the technology seems adept when comparing groups of patients, the consequences for an individual patient are unknown. Because of the limited understanding regarding the variability and inherent noise level, the robustness of the ability to detect minor changes in expression level is uncertain. Array technology has not been utilized in prospective clinical trials, and until such time as it is, the utility of microarray technology in clinical trials is difficult to determine.

III. PHARMACOGENOMIC BIOMARKERS

A. Genetic Polymorphisms Predicting Drug Metabolism

Currently the most common pharmacogenomic biomarkers utilized in drug development are the genetic polymorphisms present in the metabolic enzymes of the liver. A variety of mutations are found in the metabolizing enzymes, with single nucleotide polymorphisms (SNPs) the most common [40]. Table 2 lists the known polymorphisms of enzymes with metabolic activity. Unlike disease-associated mutations, the genetic changes in these enzymes do not predict disease or response to therapy, but lead to the absence of, or marked decrease in, metabolic activity and clearance of drugs. The list of drugs metabolized by each enzyme system is large, and reviewed elsewhere [41–43]. These enzymes are not targets for new therapeutics. Rather, the presence of a function-altering mutation

Table 2 Metabolic Enzymes of the Liver

I. Cytochrome P450 enzymes involved in drug metabolism with known genetic Polymorphisms
CYP3A1
CYP2C9
CYP2C19
CYP2E1
CYP2A6
CYP2D6
II. Other drug-metabolizing enzymes with known genetic polymorphisms: Glutathione S-transferases
GSTM1
GSTT1
N-Acetyltransferases
NAT2

in one of these enzymes often necessitates dosing regimen alterations for therapeutics metabolized by the enzyme.

For example, one of the cytochrome P450 enzymes, named 2D6 (CYP2D6), is responsible for metabolizing approximately 25% of the current cadre of commercially available drugs [44]. At the time of this writing, this enzyme has 37 known mutations of which six have been shown to have no enzyme activity [CYP2D6*3, *4, *5, *6, *9, *21, see Ref. [45]] and another two or three have been shown to have decreased activity (CYP2D6*10, *17). A patient with two copies of defective gene (e.g., homozygous CYP2D6*4 or CYP2D6*4, *6) has significantly reduced clearance of the parent drug, resulting in a prolonged half-life. The importance of reduced drug clearance is heightened when the therapeutic margin of safety is relatively low. Therefore, in individuals with decreased clearance of the drug due to mutations in a metabolic enzyme, a reduced dose is warranted. Even though these tests are not widely used outside of clinical trials, there is growing acceptance for their use in determining proper dosing regimens.

B. Genetic Polymorphisms Predicting Drug Response

In addition to polymorphisms in the genes encoding drug-metabolizing enzymes, genetic variants in the genes involved in the therapeutic pathway or promoters of these genes may influence clinical response to treatment. Therefore, genetic biomarkers may facilitate classification of individuals by level of response, improving therapeutic outcome and allowing for personalized prescriptions. Although the widespread clinical utility of such genetic biomarkers is yet to be proven, recently the literature has contained of few reports of the first step, the association of genetic variants with clinical response.

One example is the relationship between polymorphisms in the B_2-adrenergic gene and response to B agonists aimed at reversing acute bronchospasm in asthma. Even though the SNPs identified in the B_2-adrenergic receptor gene have demonstrated functional consequences in vitro and in vivo, their relationship with bronchodilatory response to B agonists remains uncertain. A multitude of reasons could account for the conflicting results including that a haplotype, or combination of SNPs, rather than an individual variant is associated with clinical response. In a recent trial 13 SNPs in the gene were organized into 12 haplotypes estimated using phylogenetic analysis. Some of these haplotypes but not individual SNPs were found to be associated with bronchodilatory response to albuterol in a sample of Caucasians [46]. The association of genetic variants with dose response is complex. Multiple SNPs within a haplotype may have a biological effect through interactions involving transcription, translation, and protein processing that ultimately affects therapeutic phenotype. As illustrated with *CAPN10*, the association of an estimated haplotype in a relatively small population is just the initial step in a long process. Thus, the association of the B_2-adrenergic receptor gene needs to be replicated in a larger sample. Furthermore, the haplotypes in the B_2-adrenergic receptor gene exhibit divergence in their frequency in Caucasian, African-American, Asian, and Hispanic Latino populations. Therefore, the association needs to be investigated in a variety of ethnic groups.

Genetic variants in the regulatory regions of genes rather than in the coding regions of the genes in the therapeutic pathway itself may influence therapeutic response. The association of the 5-lipoxygenase (*ALOX5*) promoter genotype with response to antiasthma treatment [47] provides evidence for the importance of the regulatory region. DNA sequence variants in the promoter of *ALOX5* were associated with diminished promoter reporter activity in tissue culture. Using clinical trial results for *ABT-761*, a selective inhibitor of *ALOX5*, individuals homozygous for mutant alleles demonstrated significantly decreased response as measured by FEV_1 when compared with individuals heterozygous and homozygous for the wild-type allele. Carriers of the mutant genotype may explain a portion of the individuals who do not respond to *ALOX5* inhibition but the *ALOX5* variants are not yet useful as a biomarker for routine use. The report by Drazen et al. included only 10 individuals homozygous for the mutant allele.[47] Prior to implementation of routine genetic testing of the *ALOX5* promoter in the clinic, the results of this study need to be duplicated in a larger trial with multiple outcome measures. In addition, as approximately only 6% of asthma patients do not carry a wild-type allele at the *ALOX5* promoter locus, there may be other genetic defects in the pathway yet to be identified. These examples illustrate the potential applicability of genetic biomarkers for response;

however, as with the other genetic markers discussed in this chapter, these markers are in the early stages of development.

C. Predicting Adverse Events or Side Effects

Adverse events related to drug therapy are not uncommon, causing significant morbidity and mortality [48]. Fortunately, most are mild and do not require cessation of therapy. More serious adverse events are responsible for the discontinuation of drugs during development, and in several instances even after launch. Pharmaceutical companies, and medicine in general, would be greatly served if those patients likely to experience an adverse event could be predicted. For purposes of discussion, this section will be limited to specific adverse events related to genetic mutations in genes not involved in drug metabolism. While toxicity related to altered pharmacokinetics produces adverse events, these events are handled earlier in this chapter.

Touted as one area of significant potential impact is the role of pharmacogenomics in predicting those patients likely to suffer a drug-induced adverse reaction. Unfortunately, aside from those genetic differences predicting altered drug metabolism and the associated drug toxicity, no examples exist where pharmacogenomics has predicted these unwanted occurrences. Nonetheless, genetics plays a large role in patients' reactions to medications, and pharmacogenomics offers great potential to avert unwanted side effects. In our estimation, at least four criteria need to be present for a genetic biomarker to have utility in predicting adverse events. The criteria are: (1) the adverse event must be relatively frequent, (2) the adverse event must be non-life-threatening, (3) the adverse event must be less or equal in severity to the medical condition requiring treatment, and (4) the therapy must fill a niche in the marketplace (for example, few alternatives for the therapy are available or a compelling reason to take the medication exists, such as a better formulation or improved efficacy). These conditions are especially true for the first drug to place genetic testing in its label as the current public debate surrounding the ethical and privacy issues further complicates the matter.

A recent example of a drug taken off the market following adverse-event occurrence that will be used to highlight the reasoning behind these requirements is troglitazone. Troglitazone is a thiazolidinedione antidiabetic agent for the management of type II diabetes mellitus. During clinical development, liver toxicity was noted in 48 of approximately 2500 patients, with 20 of the 48 patients withdrawing from treatment [49]. Elevation of liver enzymes was reversible on therapy cessation, and the FDA subsequently licensed the drug. During the first 2 years after launch, $\sim 1,000,000$ patients were placed on troglitazone. Of these, 70 experienced liver failure including 60 deaths and 10 transplants, leading to troglitazone's recall [49].

The frequency of the adverse event impacts the ability to associate a genetic marker with the event occurrence. In the troglitazone example, 70 out of ~ 1,000,000 patients had severe liver toxicity. Let's assume for this argument that this was a non-life-threatening event, and that all patients were willing to give a DNA sample. Even if 70 individuals sufficiently powers the study for the initial association, to replicate the results and establish a link between the polymorphism and the adverse event, an additional ~ 1,000,000 patients must evaluated to get the confirmatory sample of 70 individuals. The ethical considerations, safety of the patients, and exorbitant cost make it unlikely that any pharmaceutical company would support this effort.

For pharmacogenomics to have an impact in the instance of a life-threatening event, a sufficient number of patients need to experience the adverse event to power the analysis definitively associating the genetic biomarker with the event. Owing to rigorous monitoring of drug-related events by companies and regulatory agencies, the criteria for recall due to serious side effects, such as death, are invoked fortunately long before sufficient numbers of patients exist to prove causation. The sample size is further limited by the availability of the DNA from the individuals who experienced the adverse event. If the event is death, DNA may not be available from the individuals who experienced the event. Additionally, the survivors of a severe event may be unwilling to cooperate with the drug manufacturer. For troglitazone, a sample of 70 patients with severe liver disease is likely not sufficient to prove association with even a single genetic polymorphism. If the liver toxicity associated with troglitazone use is multifactorial with more than one polymorphism or environmental factor involved, then the 70 patients provide even less power. With a severe adverse event, ethical and liability considerations prevent the gathering of sufficient data for the subsequent development of a validated genetic biomarker.

To highlight the third and fourth considerations, patients would be unlikely to agree to troglitazone therapy, considering the severity of acute liver toxicity and available alternative therapies. Alternatives in the same class of treatment with similar pharmacokinectics without the same degree of toxicity are available. Without a compelling reason to pursue the long, expensive investigation of the relationship between an adverse event such as severe liver toxicity and genetic polymorphisms, it is unlikely that pharmacogenomics will save troglitazone or any other compound with similar problems. If the adverse event is less severe or the medical condition more severe, or in the case of a therapeutic with no or few available alternatives, pharmacogenomics may have more widespread utility. The exact frequency of events amenable for pharmacogenomic analysis will be debated, and determined on a case-by-case basis, taking into consideration the severity of the event, size of the market, and available alternatives.

Greater potential utility for genetic markers in adverse event prediction may be in the preclinical arena. Animal models have been helpful in triaging drugs that are clearly toxic to the liver, kidney, and heart. However, these models do not accurately predict all toxicity seen in humans. If correlation between animal models and human outcomes can be established, especially at the transcriptional level, the behavior of new drugs in the appropriate animal model could dictate the fate of future development. Ethical and legal considerations do not limit adverse event generation in animal models, and predicting subclinical liver, kidney, or heart toxicity in a preclinical setting could save considerable time and money by preventing failed human clinical trials. While no examples are clearly evident yet, the potential in preventing human morbidity is worth a considerable investment.

IV. DEVELOPMENT OF A GENETIC BIOMARKER ASSAY: THE HERCEPTIN STORY

An example of the widespread use of genetic biomarkers in general practice is the example of the marker used to identify breast cancer patients eligible for treatment with Herceptin. *HER2* is the protein product of the oncogene *erbB-2*, a member of the growth factor oncogene family of receptors [50]. The overexpression of *HER2* on the surface of some breast cancers is closely linked to amplification of the *erbB-2* DNA and contributes to the growth characteristics of these tumors. *HER2* overexpression has been associated in several studies with a poorer prognosis including decreased long-term survival and shorter time to relapse [51]. Shortly after the illumination of this relationship with prognosis, methods to detect DNA amplification of *HER2* including immunohistochemistry (IHC) and fluorescence in situ hybridization (FISH) analysis were developed. IHC detects an abundance of *HER2* protein on the surface of cancerous tissue, while FISH detects the amplification of *erbB-2* DNA [52].

The development of reliable methods to detect *HER2* overexpression was driven by the development of treatment-based approaches to counteract the growth stimulation signal of *HER2*. Genentech developed Herceptin, a humanized trastuzumab antibody against *HER2*, for the treatment of *HER2* overexpressing breast cancer, approximately one-third of breast cancer cases [53]. Treatment with Herceptin, especially in combination with other chemo-therapeutic agents, is now considered the standard of care for *HER2*-overexpressing breast cancers.

An assay for *HER2* was necessary to identify patients eligible for treatment with Herceptin. FDA approval came first for an immunohistochemistry-based assay (IHC), Hercep Test (Dako). Although some controversy has arisen regarding the use of IHC, the Hercep Test remains the most commonly used method for selecting patients for trastuzumab therapy [52,54,55]. Some support FISH as the clinical assay of choice stating that it has more specificity for *erbB-2* amplification and superior correlation with clinical outcomes. Varying opinions exist from experts on both sides of the fence with some going as far as to state that they believe "the FDA approved an invalid method" [55] and others endorsing the Hercep Test as very standardized with acceptable results. Although both methods are widely available and utilized, FISH is not nearly as simple to perform and is more costly than IHC. Consequently, for economic reasons, several investigators suggest screening with IHC first, followed by the more expensive FISH analysis only to resolve equivocal results. Controversy remains even with this tiered approach as some physicians believe too many patients who could benefit from Herceptin are missed using the IHC as the first line. The data simply do not exist to recommend a single method. The answer will come from trials currently underway comparing assays to treatment response. Second-generation FISH assays under development, which do not require fluorescence, may also be a viable alternative in the near future.

The development of Herceptin and the validation of a biomarker for *HER2* or *erbB-2* overexpression provide many lessons for the expectations of future pharmacogenomic markers. The role of *HER2* was first uncovered nearly 15 years ago. The generation of sufficient data supporting its role in cancer, the development of the antibody, the development of methods with adequate sensitivity and specificity to measure *HER2* expression, plus data sufficient enough for validation, took many years. A reasonable assumption for future pharmacogenomic biomarkers is that a few years can be shaved off this time line. The time necessary to link a genomic marker, or RNA level, to a specific disease state with more than one scientific analysis (Fig. 1, steps A and B) will take 1–3 years at a minimum. To proceed to generalized use, and convince the appropriate regulatory agencies, the science supporting the rationale as well as the methodology must be sufficiently compelling. In reality, this usually means more than one study linking the marker to the specific disease state, and more than one study confirming the sensitivity and specificity of the testing methodology. Multiple studies means multiple years. The idea that we will have an explosion of pharmacogenomics markers suitable for general use in the immediate future, which in the United States requires FDA approval, appears unrealistic.

V. ETHICAL CONSIDERATIONS

Other issues that need to be addressed prior to the general application of genetic markers in the clinical setting include issues related to privacy and the ethical and legal use of genetic data.[56] In a scientific research setting, addressing the difficult genetic privacy concerns may be avoided as the research may proceed without having to retain identifiable links to the individual. However, prior to widespread testing in the general clinical setting, patients and practitioners alike need to be educated regarding genetic and biological markers, the applicability of the population genetics and population data, and the importance of environmental factors. In addition, the public concerns regarding privacy and confidentiality, the potential for genetic discrimination, and the societal implications for increased information about human variation need to be addressed [56]. Without laws governing genetic privacy, individuals considering genetic testing will need to consider the risk that their personal genetic profile will become a matter of public record. Genetics will provide valuable information to assist in the management of patients and has the potential to revolutionize medicine. However, as we have attempted to outline in this chapter, the development and application of genetic biomarkers is an involved process and we are just at the beginning of a long journey toward improved therapeutic outcome.

VI. SUMMARY

Genomics and specifically genetic biomarkers hold great promise for the future of medicine. The identification of genetic markers of disease susceptibility, tumor classification, metabolism, dose response, and the development of adverse events have the potential to improve therapeutic outcome. However, prior to widespread clinical application of a genetic biomarker, multiple scientific and clinical studies must be completed to identify the genetic variants and delineate their functional significance in the pathophysiology of a carefully defined phenotype. The utility of the genetic marker, including applicability in a variety of populations, the relative risk associated with carrying the polymorphism, the population attributable risk, and what information is conferred by the test result, must be clearly understood. Subsequently, a validated assay with understood specificity and sensitivity that is available for general use must be developed. The examples outlined in this chapter are related to the general utility of the biomarker not the biomarker's specific reliability, or the degree to which the results obtained by a measurement procedure can be replicated, and the biomarker's validity, or the extent to which a biomarker measures what it is intended to measure. The statistical issues related to biomarker development are well covered in other chapters. Any one individual genetic biomarker may have the most utility when used in conjunction with a variety of other factors such as other genetic biomarkers, environmental factors, protein markers, and others.

REFERENCES

1. Lander, E.S.; Linton, L.M.; Birren, B.; Nusbaum, C. Initial sequencing and analysis of the human genome. Nature **2001**, *409*, 860–921.
2. The human genome: science genome map. Science **2001**, *291*, 1218.
3. Collins, F.S.; McKusick, V.A. Implications of the human genome project for medical science. J. Am. Med. Assoc. **2001**, *285*, 540–544.
4. Canning, D.M.; Huntsman, R.G. An assessment of Sickledex as an alternative to the sickling test. J. Clin. Pathol. **1970**, *23*, 736–737.
5. Cunningham, G.C. Phenylketonuria testing—its role in pediatrics and public health. CRC Crit. Rev. Clin. Lab. Sci. **1971**, *2*, 45–101.
6. HDCR Group. A novel gene containing a trinucleotide repeat that is expanded and unstable on Huntington's disease chromosomes. Cell **1993**, *26*, 971–983.
7. Davies, K. Genetic screening for cystic fibrosis. Nature **1992**, *357*, 425.
8. Highsmith, W.E.; Chong, G.L.; Orr, H.T.; Perry, T.R.; Schald, D.; Farber, R.; Wagner, K.; Knowles, M.R.; Warwick, W.J.; Silverman, L.M. Frequency of the delta 506 mutation and correlation with XV.2c/KM-19 haplotypes in an American population of cystic fibrosis patients: results of a collaborative study. Clin. Chem. **1990**, *36*, 1741–1746.
9. *GeneTests*; National Institutes of Health: Bethesda, MD, 2001.
10. Wilfond, B.S. Screening policy for cystic fibrosis: the role of evidence. Hastings Cent. Rep. **1995**, *25*, S21–S23.
11. Risch, N.; Merikangas, K. The future of genetics of complex diseases. Science **1996**, *273*, 1516–1517.
12. Merikangas, K.R.; Swendsen, J.D. Genetic epidemiology of psychiatric disorders. Epidemiol. Rev. **1997**, *19*, 144–155.
13. Rao, D.C.; Province, M.A. *Genetic Dissection of Complex Traits*; Academic Press: New York, 2001.
14. Kruglyak, L. Prospect for whole genome linkage disequilibrium mapping of common disease genes. Nature **1999**, *22*, 139–144.
15. Kaplan, N.; Morris, R. Issues concerning association studies for fine mapping a susceptibility gene for a complex disease. Genet. Epidemiol. **2001**, *20*, 423–457.
16. Niculescu, A.B.; Segal, D.S.; Kuczenski, R.; Barrett, T.; Hauger, R.L.; Kelsoe, J.R. Identifying a series of candidate genes for mania and psychosis: a convergent functional genomics approach. Physiol. Genomics **2000**, *4*, 83–91.
17. Horikawa, Y.; Oda, N.; Cox, N.J.; Li, X.; Orho-Melander, M.; Hara, M.; Honikio, Y.; Linder, T.; Mashima, H.; Schwarz, P.; del Bosque-Plata, L.; Yoshiuchi, I.; Colilla, S.; Polonsky, K.; Wei, S.; Cancannon, P.; Iwaski, N.; Schulze, J.; Baier, L.; Bogardus, C.; Groop, L.; Boerwinkle, E.; Hanis, C.; Bell, G. Genetic variation in the gene encoding calpain-10 is associated with type 2 diabetes mellitus. Nat. Genet. **2000**, *29*, 163–175.
18. Baier, L.J.; Permana, P.A.; Yang, X.; Pratley, R.E.; Hanson, R.L.; Shen, G.Q.; Mott, D.; Knowler, W.C.; Cox, N.J.; Horikawa, Y.; Oda, N.; Bell, G.I.; Bogardus, C. A calpain-10 gene polymorphism is associated with reduced muscle mRNA levels and insulin resistance. J. Clin. Investig. **2000**, *106*, R69–R73.

19. Evans, J.C.; et al. Studies of association between the gene for calpain-10 and type II diabetes mellitus in the United Kingdom. Am. J. Hum. Genet. **2001**, *69*, 544–552.

20. Barroso, I.; Gurnell, M.; Crowley, V.E.; Agostini, M.; Schwabe, J.W.; Soos, M.A.; Maslen, G.L.; Williams, T.D.; Lewis, H.; Schafer, A.J.; Chatterjee, V.K.; O'Rahilly, S. Dominant negative mutations in human PPARgamma associated with severe insulin resistance, diabetes mellitus and hypertension. Nature **1999**, *402*, 880–883.

21. Altshuler, D.; Hirschhorn, J.N.; Klannemark, M.; Lindgren, C.M.; Vohl, M.C.; Nemesh, J.; Lane, C.R.; Schaffner, S.F.; Bolk, S.; Brewer, C.; Tuomi, T.; Gaudet, D.; Hudson, T.J.; Daly, M.; Groop, L.; Lander, E.S. The common PPARgamma Pro12Ala polymorphism is associated with decreased risk of type 2 diabetes. Nat. Genet. **2000**, *26*, 76–80.

22. Cruts, M.; Hendricks, L.; Van Broeckhoven, C. The presenilin genes: a new gene family involved in Alzheimer disease pathology. Hum. Mol. Genet. **1996**, *S5*, 1449–1455.

23. Corder, E.H.; Saunders, A.M.; Strittmatter, W.J.; Schmechel, D.E.; Gaskell, P.C.; Small, G.W.; Roses, A.D.; Haines, J.L.; Pericak-Vance, M.A. Gene dose of apolipoprotein E type 4 allele and the risk of Alzheimer's disease in late onset families. Science **1993**, *261*, 921–923.

24. Lengauer, C.; Kinzler, K.W.; Vogelstein, B. Genetic instabilities in human cancers. Nature **1998**, *396*, 643–649.

25. Cornelisse, C.J.; Cornelis, R.S.; Devilee, P. Genes responsible for familial breast cancer. Pathol. Res. Pract. **1996**, *192*, 684–693.

26. Kinzler, K.W.; Nilbert, M.C.; Su, L.K.; Vogelstein, B.; Bryan, T.M.; Levy, D.B.; Smith, K.J.; Preisinger, A.C.; Hedge, P.; McKechnie, D. Identification of *FAP* locus genes from chromosome 5q21. Science **1991**, *253*, 661–665.

27. Groden, J.; Thliveris, A.; Samowitz, W.; Carlson, M.; Gelbert, L.; Albertsen, H.; Joslyn, G.; Stevens, J.; Spirio, L.; Robertson, M. Identification and characterization of the familial adenomatous polyposis coli gene. Cell **1991**, *66*, 589–600.

28. Cho, K.R.; Vogelstein, B. Suppressor gene alterations in the colorectal adenoma-carcinoma sequence. J. Cell Biochem. **1992**, *S 16G*, 137–141.

29. Fearon, E.R.; Vogelstein, B. A genetic model for colorectal tumorigenesis. Cell **1990**, *61*, 759–767.

30. DeMali, K.A.; Godwin, S.L.; Soltoff, S.P.; Kazlauskas, A. Multiple roles for *Src* in a PDGF stimulated cell. Exp. Cell Res. **1999**, *25*, 271–279.

31. Nebigil, C.G.; Launay, J.; Hickel, P.; Tournois, C.; Maroteaux, L. 5-Hydroxytryptamine 2B receptor regulates cell cycle progression: cross talk with tyrosine kinase pathways. Proc. Natl Acad. Sci. U.S.A. **2000**, *97*, 2591–2596.

32. Kalidas, M.; Kantarjian, H.; Talpaz, M. Chronic myelogenous leukemia. J. Am. Med. Assoc. **2001**, *286*, 895–898.

33. Listed, N. Gleevec (STI-571) for chronic myeloid leukemia. Med. Lett. Drugs Ther. **2001**, *43*, 49–50.

34. Druker, B.J.; Lydon, N.B. Lessons learned from the development of an abl tyrosine kinase inhibitor for chronic myelogenous leukemia. J. Clin. Invest. **2000**, *105*, 3–7.

35. Schulze, A.; Downward, J. Navigating gene expression using microarrays—a technology review. Nat. Cell Biol. **2001**, *3*, E190–E195.

36. Alizadeh, A.A.; Eisen, M.B.; Davis, R.E.; Lossos, I.S.; Rosenwald, A.; Boldrick, J.C.; Sabet, H.; Tran, T.; Yu, X.; Powell, J.I.; Yang, L.; Marti, G.E.; Moore, T.; Hudson, J.; Lu, J.; Lewis, D.B.; Tibshirani, R.; Sherlock, G.; Chan, W.C.; Greiner, T.C.; Weisenburger, D.D.; Armitage, J.O.; Warnke, R.; Staudt, L.M. Distinct types of diffuse large B-cell lymphoma identified by gene expression profiling. Nature **2000**, *403*, 503–511.

37. Khan, J.; Wei, J.S.; Ringer, M.; Saal, L.H.; Ladanyi, M.; Westermann, F.; Berthold, F.; Schwab, M.; Antonescu, C.R.; Peterson, C.; Meltzer, P.S. Classification and diagnostic prediction of cancers using gene expression profiling and artificial neural networks. Nat. Med. **2001**, *7*, 673–679.

38. Alaiya, A.A.; Franzen, B.; Hagman, A.; Silfversward, C.; Moberger, B.; Linder, S.; Auer, G. Classification of human ovarian tumors using multivariate data analysis of polypeptide expression patterns. Int. J. Cancer **2000**, *86*, 731–736.

39. Sorlie, T.; Perou, C.M.; Tibshirani, R.; Aas, T.; Geisler, S.; Johnsen, H.; Hastie, T.; Eisen, M.B.; van de Rijn, M.; Jeffrey, S.S.; Thorsen, T.; Quist, H.; Matese, J.C.; Brown, P.O.; Botstein, D.; Eystein-Lonning, P.; Borresen, D. Gene expression patterns of breast carcinomas distinguish tumor subclasses with clinical implications. Proc. Natl Acad. Sci. U.S.A. **2001**, *98*, 10869–10874.

40. Streetman, D.S.; Bertino, J.S.; Nafziger, A.N. Phenotyping of drug metabolizing enzymes in adults: a review of in vivo cytochrome P450 phenotyping probes. Pharmacogenetics **2000**, *10*, 187–216.

41. Cholerton, S.; Daly, A.K.; Idle, J.R. The role of individual human cytochromes P450 in drug metabolism and clinical response. Trends Pharmacol. Sci. **1992**, *13*, 434–439.

42. Slaughter, R.L.; Edwards, D.J. Recent advances: the cytochrome P450 enzymes. Pharmacokinetics **1995**, *29*, 619–624.

43. Spatzenegger, M.; Jaeger, W. Clinical importance of hepatic cytochrome P450 in drug metabolism. Drug Metab. Rev. **1995**, *27*, 397–417.

44. Benet, L.Z.; Kroetz, D.L.; Sheiner, L.B. Pharmacokinetics. In *Goodman, Gilman's the Pharmacologic Basis of Therapeutics*, 9th Ed.; Hardman, J.G., Gilman, A.G., Limbird, L.E., Eds.; McGraw-Hill: New York, 1996; 3–27.

45. Daly, A.K.; Brockmoller, J.; Broly, F.; Eichelbaum, M.; Evans, W.E.; Gonzalez, F.J.; Huang, J.D.; Idle, J.R.; Ingelman-Sundberg, M.; Ishizaki, T.; Jacqz-Aigrain, E.; Meyer, U.A.; Nebert, D.W.; Steen, V.M.; Wolf, C.R.; Zanger, U.M. Nomenclature for human *CYP2D6* alleles. Pharmacogenetics **1996**, *6*, 193–201.

46. Drysdale, C.M.; McGraw, D.W.; Stack, C.B.; Stephens, J.C.; Judson, R.S.; Nandabalan, K.; Arnold, K.; Ruano, G.; Liggett, S.B. Complex promoter and coding region beta 2-adrenergic receptor haplotypes alter receptor expression and predict in vivo responsiveness. Proc. Natl Acad. Sci. U.S.A. **2000**, *97*, 10483–10488.

47. Drazen, J.M.; Yandava, C.N.; Dube, L.; Szczerback, N.; Hippensteel, R.; Pillari, A.; Israel, E.; Shork, N.; Silverman, E.S.; Katz, D.A.; Drajesk, J. Pharmacogenetic association between *ALOX5* promoter genotype and the response to anti-asthma treatment. Nat. Genet. **1999**, *22*, 168–170.

48. Adverse drug events: substantial problem but magnitude uncertain. GAO/HEHS-00-21. 2000, Jan 18, 1–12.
49. Gale, E. Lessons from the glitazones: a story of drug development. Lancet **2001**, *357*, 1870–1875.
50. Wright, C.; Prasad, K.; Lennard, T.J. *ErbB-2* expression in breast and other human tumors. J. Clin. Pathol. **1992**, *45*, 459–460.
51. Piccart, M.; Lohrisch, C.; Di Leo, A.; Larsimont, D. The predictive value of *HER2* in breast cancer. Oncology **2001**, *61* (Suppl. S2), 73–82.
52. Kaptain, S.; Tan, L.K.; Chen, B. *HER2*/neu and breast cancer. Diagn. Mol. Pathol. **2001**, *10*, 139–152.
53. Cobleigh, M.A.; Vogel, C.L.; Tripathy, D. Multinational study of the efficacy and safety of humanized anti-*HER2* monoclonal antibody in women who have *HER2*-overexpressing metastatic breast cancer that has progressed after chemotherapy for metastatic disease. J. Clin. Oncol. **1999**, *17*, 2639–2648.
54. Hanna, W. Testing for *HER2* status. Oncology **2001**, *61* (Suppl. S2), 22–30.
55. Check, W. Working out the kinks in HER2 testing. CAP Today **2000**, 14.
56. Rothstein, M.A. *Genetic Secrets: Protecting Privacy and Confidentiality in the Genetic Era*; Yale University Press: New Haven, CT, 1997.

10

Quality Assurance and Regulatory Compliance

Linda F. Knoob
Eli Lilly and Company, Indianapolis, Indiana, U.S.A.

Angela Berns and Jacqueline Hevy
Amgen Inc., Thousand Oaks, California, U.S.A.

I. CHALLENGES

A. Introduction

Today, there is a continual challenge in clinical drug development to understand and implement a complete quality system that will ensure data integrity and regulatory compliance. This is especially true regarding the application of biomarkers. With the use of novel biomarkers increasing at an exponential rate, global regulations have been unable to encompass all the technologies utilized. In fact, most biomarker technologies have little or no specific regulations directing them. The goal of this chapter is to identify:

> The challenges for attaining and maintaining compliance
> The components of existing regulations that should guide the development of a quality system for diagnostic service providers (DSP).

B. Applicable Regulations

To assess the regulatory requirements for an institution such as a DSP, one must first determine whether the biomarker data will be used to support any regulated study or process. Once a regulated study has been identified, then the applicable regulations will depend on the regulatory agency/ies that will review the study report. In global submissions, the most inclusive regulations are usually emulated. In general, the global regulations for clinical drug development were adapted from the U.S. Food and Drug Administration (US FDA) regulations. The Organisation for Economic Co-operation and Development's Good Laboratory Practice (OECD GLP) [1] encompasses the US FDA's Good Laboratory Practices (GLP) [2] and the International Conference on Harmonization's Good Clinical Practice (E6) (ICH GCP) [3] encompasses the US FDA's Good Clinical Practices [4].

C. Good Laboratory Practices

If the data generated by the DSP support a preclinical animal safety study, then the Good Laboratory Practices (GLP) [1,2] will apply. GLP regulations describe the many components of a quality system that must be implemented for preclinical biomarker data generation, which include processes and documentation requirements for the following (US FDA GLP 21 CFR part 58 [2]:

> Organization/personnel (.29)
> Management (.31)
> Quality assurance unit (.35)
> Facilities (subpart C)
> Equipment (subpart D)
> Standard operating procedures (.81)
> Reporting (.185)
> Record retention/storage (.190, .195).

GLP regulations do not apply to human clinical trials but the principles of these regulations could be applied to a DSP where feasible as a means to assist in the development of a quality system.

D. Good Clinical Practices

For human clinical trials, the principles of the GCP [3,4] help describe a quality system in the clinical management of the study. There are many components of an effective quality system outlined in these regulations that would be applicable, although it may not always be apparent to DSPs that these regulations also apply to them. A first look in the GCPs where biomarkers would specifically be included is found in the International Conference on Harmonisation E6 GCP 5.18.4(b) [3], stating: "Verifying ... that facilities, including laboratories, equipment,... are adequate to safely and properly conduct the trial and remain adequate throughout the trial period." The challenge for DSPs is to determine how to become and remain "adequate." An "adequate" organization established scientifically valid biomarker techniques within a quality system that ensures data integrity, including sufficient documentation for study reconstruction. Some components of an "adequate" organization can be inferred to DSPs from the general principles of the ICH GCPs [3]:

> Education, training, and experience should qualify all individuals participating in the conduct of the study (2.8).
>
> All clinical trial information should be recorded, handled, and stored in a way that allows its accurate reporting, interpretation, and verification (2.10).
>
> Systems with procedures that assure the quality of every aspect of the trial should be implemented (2.13).
>
> Quality control should be applied to each stage of data handling to ensure that all data are reliable and have been processed correctly (5.1.3).

One needs to understand not only what regulations apply but also how they are interpreted by the regulatory agencies. In other words, what is the regulatory agency expectation for compliance? Opportunities exist to learn these expectations by reviewing additional regulatory agency documents in addition to the published regulations. These include compliance manuals, guidance documents, and inspection reports. In the United States, observations of noncompliance with regulations are documented in FDA's Form 483, Establishment Inspection reports (EIR) and Warning Letters. These documents are available from the FDA. Attending industry conferences where industry experts and regulatory agencies participate also provides opportunities for learning regulatory expectations for compliance.

E. Industry Standards

The regulatory agency expectation can be driven by industry standards. If the industry standard has defined a specific process to ensure quality, then it may be expected that all similar organizations follow this standard. For example, if specific qualifications, such as personnel education, have been deemed necessary by industry to adequately perform a task, then, although regulations in drug development do not define this specific requirement, it now becomes the regulatory expectation. Nongovernmental accrediting organizations can also define industry standards, such as the College of American Pathologists (CAP) [5] for clinical diagnostic laboratories or the general quality system outlined in the International Organization for Standardization (ISO) [6]. It is important for DSPs to understand their industry standards and how they might drive regulatory expectations. The challenge continues because industry standards are always changing. DSPs must continually evaluate their organization for compliance to the industry standards. If a biomarker technology is new, then industry standards may not be defined. The opportunity exists for new biomarker technology to establish their quality standards as the industry standard.

F.   Contract Research Organizations

When a DSP is contracted by a sponsor to support clinical trials, the responsibilities, which include the applicable quality components outlined in the GCPs, are then transferred from the sponsor to the DSP as a contract research organization (CRO). According to the ICH, consolidated GCP (E6) Guidelines [3], a CRO is "a person or an organization (commercial, academic or other), contracted by the sponsor to perform one or more of a sponsor's trial-related responsibilities." For a number of years there has been a trend in the drug development industry to outsource portions of clinical research to CROs. Indeed, the FDA issued in 1994 internal guidance on this subject as part of its Bioresearch Monitoring Program (BIMO) for inspection of "sponsors, contract research organizations and monitors" [7,9]. Many academic institutions, often attached to universities and hospitals, have also entered into the world of contract research. Unlike the larger, more established CROs, the staff at these smaller institutions generally has scientific and/or medical expertise but may lack specific training in GCP or training in the regulations surrounding the conduct of clinical research. Additionally, they may use methods that are not always validated to the extend required by the regulations. In these institutions, the staff is generally well trained in the area of research, and may even be considered experts in their particular field. However, it has been noted that these academics often enter the field of clinical research without realizing the full extent of the legal requirements. Although the sponsor might formally contract with these institutions and document transfer of obligations as required by the regulations (FDA GCP:21 CFR 312.52 and ICH GCP 5.2) [3,4], the institution may not be fully compliant with the regulations. The quality and integrity of the trial data are most important to the sponsor to ensure a successful submission. Therefore, it is essential that the sponsor play an active role in ensuring that the companies/institutions used to assist them in their research are compliant with the applicable regulations. The sponsor's regulatory compliance and quality system expectations must be communicated to the CRO and put in place. This may be directed by the study protocol or outlined in the contract, implemented through training, and should be verified by sponsor monitoring and quality assurance audits.

II.   QUALITY SYSTEMS

A.   Introduction

In the pharmaceutical industry, a quality system may be defined as a system that produces accurate, consistent, and reproducible preclinical or clinical data. The principles of a quality system should be incorporated into the development of new biomarker technology to assure preclinical and clinical researchers and the study sponsors that the data collected are acceptable to the FDA and other international regulatory agencies.

During the development of work processes, documented procedures and the assurance of a complete audit trail are often overlooked or put aside to be completed at a later date. However, for quality to exist in a process, it must be incorporated into that process during the development, and not applied as an afterthought. It is common to hear complaints from small, independent laboratories and niche provider services that the incorporation of quality processes is too time-consuming and costly. It must be understood that quality standards are required by the US Code of Federal Regulations and that clinical study participants (i.e., patients) have a right to the quality and safety standards set forth in the federal regulations and ICH guidelines. All providers of clinical data in research used to support US Investigational New Drug Applications (INDs), and subsequently any marketing applications, are held to these standards.

There are risks in not having a quality system in place and consequences associated with submitting poor quality data. A regulatory agency may reject the data produced if they are deemed inadequate or inaccurate, costing far more than the price to comply with these regulations. Losing "first to market" status can incur serious financial losses for a pharmaceutical company. A company, clinical laboratory, contract research organization, principal investigator, or DSP may be issued an FDA inspection form 483, which lists deficiencies noted during an FDA inspection. As a result of serious FDA inspection observations, a warning letter may be issued. Warning letters are available to all under the Freedom of Information Act and are published on the FDA website (www.fda.gov/foi/-warning.htm). If such a warning letter was issued to a CRO conducting research on behalf of a sponsor, there would be negative effects on the future business of that CRO as sponsors would not want to risk a delayed or rejected submission with a possible poor-quality submission. The FDA or other regulatory agency may impose fines and sanctions for noncompliance with their regulations.

B.   Components of a Quality System

A quality system should have the following components, regardless of the type of tests, analyses, or instrumentation used (Fig. 1).

Staff Qualifications and Training

Standard Operating Procedures

Documentation Standards

Appropriate Record Storage

Validation and Testing

Equipment Maintenance, Calibration and Standardization

Quality Control

Quality Assurance

Management Support

Figure 1   Components of a quality system.

1.   Staff Qualifications and Training

Documentation of staff qualifications and training is required to demonstrate that each individual involved in the conduct of a clinical trial is qualified through education, training, and experience to perform his or her respective tasks. This is usually maintained in the form of current résumés/curriculum vitae (CV), applicable training records, and appropriate licensure. The industry standard for updating résumé/CV is every 2 years. Initials and date on the résumé/CV confirms the date the document was created or updated, if the document is not dated in another way. Personnel roles and responsibilities should also be defined. This information is usually outlined in organization charts and personnel job descriptions.

2.   Standard Operating Procedures (SOPs)

Depending on the size of the company providing the biomarker data, SOPs that identify major processes may be appropriate. ICH GCP [3] defines SOPs as "detailed, written instructions to achieve uniformity of the performance of a specific function." Once SOPs are developed and put in place, they will be included in the standard by which a regulatory agency will make assessments. As procedures change over time, it may be necessary to know what the procedure was at a particular point in time. It is important, therefore, that SOPs are version-controlled documents and retained in the same manner as any study records. Documentation of training on each specific version of the SOPs when applicable

Management Control
Facilities
Equipment / Instrument Calibration and Maintenance
Materials and Reagents
Biological Sample Management
Records / Documentation Requirements, Change Controls, Archiving
Computer Systems

Figure 2 Examples of major processes.

to that person's job should be maintained. There should be a SOP on writing, distributing, changing, and retiring SOPs, i.e., an SOP on SOPs. Examples of major processes where SOPs may be required are outlined below with a description for each category (Fig. 2).

a. *Management control* includes identifying responsibilities and accountabilities, setting administrative policy and procedures, assuring adequate quality control procedures, and providing quality assurance. The institution and enforcement of SOPs are evidence of management control.

b. *Facilities* include plant maintenance, operations, physical security, emergency plans, disaster recovery including any computer systems, and safety procedures.

c. *Equipment* procedures should include inspection, maintenance and calibration (if applicable), documentation requirements, procedures for removing from service, returning to service, and permanently retiring instrumentation and machinery.

d. Standard procedures should be in place for the receipt, preparation, storage, and disposal of *materials and reagents.* Lot or batch numbers should be recorded with the assay documentation.

e. *Biological sample management* procedures include collecting, handling, storing, transporting, and archiving biological samples to ensure the integrity of the samples and the safety of those handling the samples.

f. *Records, documentation requirements, change control, and archiving* procedures may be handled as administrative procedures if standardized across all areas, or these procedures may be documented as part of other SOPs as applicable.

g. *Computer systems and electronic data management* procedures will be
described in more detail later.

3. Documentation Standards

Creating and maintaining accurate and adequate source documents are crucial to
ensure a compliant audit trail of clinical data. What is included in study records
and study data must tell a complete story to be clear to a reviewer or inspector
long after the work has been completed. Such documents may be reviewed at any
time during the lifetime of the drug; this time could be as long as 5–7 years later
when key personnel are no longer available to aid in study reconstruction. Study
documentation should include raw data, changes in personnel, methods,
instrumentation identification, or materials utilized.

Deviations to procedures should be documented with explanations and/or
corrective actions. Documentation must be legible, and corrections to the data, if
necessary, need to be made in such a way that does not obscure the original data
and indicates the reason for change [FDA GLP 58.130 (e) and ICH GCP 4.9.3]
[2,3]. This change control process is applicable to paper and electronic systems.
The industry standard for paper documents is to use one line through the error so
as not to obscure the original entry; then the correct entry is made and the
transaction is initialed and dated by the person performing the correction.
Changes made to electronic records are captured in audit trails. More information
on the regulations covering electronic records is included below.

4. Appropriate Record Storage: Short and Long Term

Study documents are proprietary and should be stored in a secure, controlled
environment. Regardless of the type of documentation provided (e.g., automated
instrument reports, written test results, films, or photographs), all study
documentation should be maintained and stored in a manner that offers protection
from degradation, unintended alterations, and damage. The stored records should
be easily and readily retrievable should a regulatory inspection occur. Fire
suppression methods should not put the data at risk. For example, paper records
should not be filed in cardboard boxes and stored under a water sprinkler fire
suppression system.

A documented system of storage and archiving procedures is
recommended. The ICH E6 GCP 3.4 [3] requires that the sponsor and
investigator retain records for a period of at least 3 years after completion of the
trial. If the data are generated by a CRO, then the sponsor or a clinical
investigator may request that records and results be maintained for that period at
the CRO.

5. Validation and Testing

Validation and testing should be performed on instruments, including computer systems, and assay methods by qualified individuals that test the limits of the system(s). Method validation should be performed on all analytical assays, which includes the specific instrumentation to be utilized, to ensure reproducibility and accuracy. This is especially important for esoteric testing or new technologies where there are no industry standards and methods. If biospecimens are utilized in the testing, then specimen stability must be determined for the storage conditions employed. Operational qualification/performance qualification (OQ/PQ) checks should be performed on instrumentation. Instruments should have routine checks for performance characteristics, blanks, controls, calibration, and maintenance at appropriate intervals. Participation in national proficiency testing programs, such as the College of American Pathologist's CAP Surveys, is recommended when applicable.

6. Equipment Maintenance, Calibration, and Standardization

There should be documentation supporting the fact that the instrument checks were performed. Logbooks or records that show appropriately timed maintenance and calibration checks should be maintained and subject to quality control procedures. Instrument downtime should be recorded and explained, along with appropriate service documentation. As with all documentation, entries need to be legible and corrections should be made in an appropriate manner, as described previously. When storage temperature is critical to ensure bioanalytical specimen integrity, storage conditions should be monitored and records should be maintained. More specific computer systems expectations are addressed below.

7. Quality Control

Quality control (QC) is the routine control of the quality of a product or a deliverable as measure against a defined standard within a company, system, or process. QC checks are measurements of quality with defined pass/fail criteria. The QC process needs to be sufficient to ensure reliable, accurate results. A statistically significant number of points must be checked often enough to provide confidence in the quality of the systems. QC checks should identify and document noncompliance with procedures and include a plan for following up and correcting those issues identified and should include appropriate management oversight. Procedures for handling system check failures should be in place.

8. Quality Assurance

A sponsor company, independent quality assurance (QA) consultants, or internal personnel not associated with the processes being audited may perform QA audits. Plans or procedures should be in place for audits by sponsor companies who routinely request visitation privileges in contracts with service providers. As a DSP, it is likely that a visit will occur if the data provided will be used in a US investigational new drug (IND) application or equivalent.

Internal audits could include regularly scheduled audits of individual systems/process, unannounced audits, and a full annual audit of all systems/processes. QA audits performed by external consultants can be valuable learning experiences for all who generate clinical data. An audit plan should identify responsibility, accountability and deadlines for audit observation closure.

A QA representative or DSP management may be called upon to host any audit. A DSP should be inspection-ready at all times whether a sponsor QA unit or a regulatory agency investigator initiates an audit. The DSP organization should identify key topic experts to be available to answer questions and organize essential quality system documents (for example, organization charts, training records, CVs, and standard operating procedures with index) to be readily retrievable at the initiation of any audit. The goal is to facilitate the conduct of the audit and to provide evidence of any effective quality system to satisfy any auditor's concerns.

9. Management Oversight and Support

To establish and maintain a compliant quality system, management must provide continual support and resources. Many quality systems are initiated and may even be completed, but so many times become disregarded by all when management puts value on speed and not quality. It is critical that management be fully committed to maintaining a compliant quality system to ensure success.

III. COMPUTER COMPLIANCE CHALLENGES

A. Introduction

Computer systems, software applications, and automatic devices play an increasingly important role in the conduct of clinical trials. Clinical research relies heavily on computer systems, for example digitized scans, automated laboratory information management systems (LIMS), and on-line data collection. However, no matter which method or scientific principle is used, there is one common denominator for all data gathered—at some stage, data will be entered into an electronic data-handling system. Once in this system, the data may be manipulated, transformed, exported, imported, analyzed, derived, or integrated into tables. The data, in final form, will be submitted to a regulatory agency for review and approval. The sponsor invests a significant amount of time working with clinical investigators and research scientists to ensure that the data are collected in a scientific manner and in accordance with a predetermined protocol. In the effort to gather scientifically sound clinical data, the integrity of the system wherein the data will reside is often overlooked. It is important to ensure that the electronic systems that store and manage these data are reliable and robust. Poorly validated systems often result in a failure to ensure the integrity of the data for the duration of its lifetime within that system. Thus, systems that support clinical data must be validated.

B. Electronic Data Management/Data Integrity

During the conduct of many clinical trials, the sponsor may outsource some or all of the research responsibilities. This is particularly true when the clinical research being conducted involves the use of specialized biomarker technologies. As a result of outsourcing, clinical trial data may reside in various systems in several organizations, for example, the sponsor, the contract research organization (CRO), the central laboratory, and the service provider performing the biomarker work. During the course of the trial, data may be transferred from one system to another system as it passes from one "owner" to the next "owner." Thus, it is necessary to ensure that all transfer processes are validated to provide assurance that the data are in no way altered or corrupted during migration from one system to the next. These complex data transfer processes can be illustrated in the following example where the biomarker technology service is an imaging center supporting a clinical trial.

Initially, these data are collected at the investigator's site and are captured and stored in the first system as an image on a disc.

The data are then transferred electronically or by hard copy to the imaging center for analysis. The data are entered, by either manual or electronic methods, into a second system, i.e., the image-handling/analyzing system.

The data are then manipulated by software and the analyzed data are stored. The data are then transferred to the central data repository (CDR), which resides at the CRO. Again, the data are exported and may be imported into the CDR manually or electronically.

For the data to be in suitable form for handover to the CDR system, they may be converted into SAS datasets or loaded into Oracle tables.

The data may then be "cleaned" (compared to a set of expected values and queried, reviewed, and corrected).

The data are then exported to the sponsor database where it may be further cleaned, analyzed, presented as listings, figures, or tables, and linked to other datasets, such as those from the central laboratory.

Finally, the data are configured into an electronic form acceptable to the regulatory agency and are submitted on electronic media.

The number of transactions, transfers, and manipulations that these data have undergone can become quite complicated. The data may have been generated years before it is finally submitted for review. In addition, the image gatherers, the image readers, the analyzers, the investigator, the study nurse, the data monitor, the safety monitors, the statisticians, and the medical writers may all handle these data. Assuming these people are qualified by training and experience and thus know how to manage the data, one then needs to be assured that all the systems that are used to process, manipulate, and transfer these data are validated and functioning as required.

C. Regulations and Computer Systems

The production of scientifically sound data stored in robust computer systems is not just desirable; it is the law. Depending on the nature of the investigative product, computer systems used in clinical trails are subject to the requirements of the GCP, GLP, or Good Manufacturing Practices (GMP) regulations. Thus, all data included as part of a regulatory submission for approval are covered by regulations requiring documented evidence of data integrity and validation of the data-handling systems. Good validation practices and principles have long been established and practiced in the pharmaceutical industry. The US FDA provides a collection of regulations for GCP in its Code of Federal Regulations, including 21 CFR Part 312, Investigational New Drug Application [4]. The guide to inspection of computerized systems in drug processing, more commonly known as the Blue Book, was published in 1983 [8]. In 1987, the FDA issued a Technical Reference on Software Development Activities, which functioned as a reference and guide for inspectors [9]. The FDA extended the regulations in 1997 with the implementation of the Electronic Records and Signature Rule, 21 CFT Part 11 [10]. When electronic records and/or electronic signatures are being used, 21 CFR Part 11 [10] is applicable to all FDA program areas. Thus, any system that generates or maintains electronic records for compliance with FDA regulations (including predicate rules) is subject to inspection for compliance with 21 CFR Part 11 [10]. Further guidance for industry was provided in April 1999 with the publication of a guideline entitled "Computerized Systems used in Clinical Trials" [11]. In May 1999, the FDA issued the compliance policy guide "Enforcement Policy: 21 CFR Part 11: Electronic Records; Electronic Signatures" (CPG 7153.17) [12].

D. Compliant Computer System

There are several areas that the researcher/sponsor should consider to ensure that biomarker data are captured in electronic systems compliant with regulations. Many systems used in clinical trials are purchased as commercial off-the-shelf (COTS) packages. However, owing to the nature and variety of technical requirements in the clinical research arena, many systems are either custom-built or are user-configured. As such, each system must be examined as a separate entity and evaluated for compliance with the applicable regulations. This section will present the elements of a compliant system including those described in 21 CFR Part 11 [10]. It is not intended that this be an all-inclusive analysis of the rule. Rather, it is intended to highlight the parts of the rule that are reflected in a quality compliant system.

1. Open and Closed Systems

In CFR 21 Part 11 [10], electronic systems are defined as "open" or "closed." A closed system is one in which access to the system is controlled by persons who are responsible for the content of electronic records that are on the system. An open system, on the other hand, is an environment where system access is not controlled by persons for the content of electronic records that are on the system. The FDA has identified requirements for the use of closed systems used to create, modify, maintain, or transmit electronic records. Such controls include the use of tools designed to ensure the authenticity, integrity, and confidentiality of electronic records from the point of their creation to the point of their receipt. When electronic signatures are being used, controls need to be in place to ensure that the signer cannot readily repudiate the signed records as not genuine. Persons who use open systems to create, modify, maintain, or transmit electronic records must employ similar procedures and controls as for closed systems and include additional security features. Such additional features may include document encryption and use of appropriate digital signature standards to ensure record authenticity, integrity, and confidentiality.

FDA's 21 CFR Part 11 [10] regulations present a majority of elements that are required in any system that claims to have a robust quality management system in place. The procedures and controls required by these regulations for a closed system include the following:

a. Validation of systems to ensure accuracy, reliability, consistent intended performance, and the ability to discern invalid or altered records.
b. The ability to generate accurate and complete copies of records in both human-readable and electronic form suitable for inspection, review, and copying by the regulatory agency.

c. Protection of records to enable their accurate and ready retrieval throughout the records retention period.
d. Limiting system access to authorized individuals.
e. Use of secure, computer-generated, time-stamped audit trails.
f. Use of operational system checks to enforce permitted sequencing steps and events, as appropriate.
g. Use of authority checks to ensure that only authorized individuals can use the system, electronically sign a record, access the operation or computer system input or output device, alter a record, or perform the operation at hand.
h. Use of device (e.g., terminal) checks to determine, as appropriate, the validity of the source of data input or operational instruction.
i. Determination that those persons, who develop, maintain, or use electronic record/electronic signature systems have the education, training, and experience to perform their assigned tasks.
j. The establishment of, and adherence to, written policies that hold individuals accountable and responsible for actions initiated under their electronic signatures, to deter record and signature falsificaiton.
k. Use of appropriate controls over systems documentation including:

1. Adequate controls over the distribution of, access to, and use of documentation for system operation and maintenance.
2. Revision and change control procedures to maintain an audit trail that documents time-sequenced development and modification of systems.

2. Validation

When validated electronic systems are used, one can be assured that the quality of the data can be supported at every stage of data processing and during every data transfer. A validated system is one that has addressed data integrity, system reliability and reproducibility, change control, security, and supporting documentation. If the system is custom-built or a configured COTS, the system should be validated for intended use by following an established system development life cycle (SDLC). This life cycle established the procedures to be followed, and documentation to be maintained, from the concept phase to retirement of the system. Evidence that the process has been followed should be carefully documented, filed, and ready for inspection. These documents may include user requirements, design specifications, a validation protocol, test plans, scripts and results, validation summary, qualification documentation [installation qualification (IQ), operational qualification (OQ), and performance qualification (PQ)], and the retirement procedure. Such documented evidence of the validation effort may be used to demonstrate to sponsor auditors and regulatory agencies that the system is in compliance with the applicable regulations.

Documented change control methods should be established to ensure that changes, such as software upgrades or the application of patches to fix bugs, to a validated system are recorded. Obviously, such changes should also be tested and validated according to the SDLC. Validation efforts should be conducted in parallel with system development. Legacy GxP systems that were in existence before FDA's 21 CFR Part 11 regulation [10] may still be operational without being complaint with the regulation. Retrospective validation of such systems is required but is a tricky exercise and should be conducted in consultation with a validation expert.

A compliant computer system should be:

Located in a secure, controlled environment with appropriate fire suppression

Installed to the manufacturer's directions (installation qualification)

Tested to ensure that each unit of the system operates as intended (operational qualification)

Tested for expected performance in the normal operating environment (performance qualification)

Supported by SOPs

Used by personnel who are trained in the use of the system and the associated SOPs

In addition to validation requirements, other elements of a compliant computer system covered by FDA's 21 CFR Part 11 [10] are listed below.

3. Electronic Audit Trails

Audit trails are a requirement of FDA's 21 CFR Part 11[10] and ICH GCP (5.5.3). [3]. The audit trail should capture all operator entries and actions that create, modify, or delete records. The audit trail should record the name (or user identification) of the person making the change, the date and time of the change, the operation performed (change, deletion, or modification), and the reason for change (if required by the predicate rule). A compliant system should include the following attributes:

The ability to generate secure, computer-generated, time-stamped entries that independently record the date and time of operator entries and actions that create, modify, or delete electronic records

The ability to discern invalid or altered records in a database

The audit trail mechanism in the validation of the electronic records system

The following is also required of audit trails:

Changes made to records should not obscure previously recorded entries.
The audit trail should be retained for at least as long as the associated electronic records.
The audit trail should be available for agency review and copying.
The mechanism for retrieving the audit trail should be validated.

4. Electronic Security

Authority checks are a requirement of FDA's 21 CFR Part 11 [10]. The system should be designed to limit access such that only authorized individuals can use the system, electronically sign a record, alter a record, or perform an operation. These authority checks also need to control the level of access to the system. Users should only have access appropriate to their level of training and experience. For example, a person new to using a system should not be given administrator access.

5. Regulatory Inspection for FDA's 21 CFR Part 11 Compliance

FDA inspectors can conduct an audit against the requirements for 21 CFR Part 11 [10] when they are conducting a sponsor audit. Whether working in a GCP, GLP, or GMP environment, if the inspector is reviewing a system, process, or documentation that falls under the predicate rules, he may review the system supporting the process for compliance with 21 CFR Part 11 [10]. The enforcement actions for 21 CFR Part 11 [10] deviations will depend on the nature and extent of the deviations, the effect on the product quality and data integrity, and the compliance history of the company being inspected. Warning letters referencing 21 CFR Part 11 [10] are available on the FDA website at http://www.fda.gov/foi/warning.htm. Some of the key issues noted during the agency's inspections are listed below:

Validation problems.
Inadequate documentation.
Programs and macros not validated.
Failure to keep up with standards and enabling technologies, for example, use of databases with noncompliant audit trails.
Inadequate change control and configuration management for remote changes by vendor and interface changes.
Poor network security.
Older systems are less secure than traditional paper.
Record-integrity principles and practices are outdated.
All users have system administrator privileges.
Network administrator unqualified.

Passwords posted to directory.

Common passwords that lessen security, such as "password" or "12345."

E. Internet Challenges

A recent technological advance is having a huge impact on clinical research—the Internet. The latest hardware and software, such as remote data entry systems and electronic case report forms accessible over the world wide web, are displayed proudly at trade shows and seminars. These technologies promise to cut the time to approval and thus have captured the attention of many sponsors who are only too aware of the millions of dollars lost every day a submission is delayed. A new CRO is evolving—the data management organization (DMO). These vendors use web-based browsers to collect and manage clinical trial data. Investigator sites participating in web-based trials can access the CDR at the DMO via a computer and a web browser. This opens the possibility of activating thousands of sites at once, sending instant updates, messages, and data queries at the click of a button. Of course, with new technology come new challenges. The new methodology introduces concerns over protected access to the database, change control issues, security, and electronic signatures. One should refer to the section on auditing of CROs and ensure that CROs supplying such services are fully aware of, and compliant with, the requirements of 21 CFR Part 11 [10].

IV. SUMMARY

The challenge to develop and sustain a quality system will continue as biomarker technology evolves and regulatory expectations change. It is important to continually monitor the regulatory environment for changes in compliance expectations. DSPs have an enormous challenge to meet. As well as ensuring the validity of the scientific methods employed in the generation of clinical data, DSPs must also strive to ensure that their operations are compliant with applicable regulations. Even though sponsors outsource biomarker data generation to CROs or academic institutions, the ultimate responsibility for the accuracy and integrity of the clinical data and the systems used to handle these data lies with the sponsor. As such, sponsors should exercise due diligence to ensure that the data providers operate in compliance with the regulations and are aware of the principles of quality management system that ultimately leads to a successful regulatory marketing application approval.

V. GLOSSARY

A. Audit

A systematic and independent evaluation of trial-related activities and documents to determine whether the evaluated trial-related activities were conducted, and the data recorded, analyzed, and accurately reported according to the protocol, sponsor's SOPs, GCP and applicable regulatory requirement(s) [3].

B. Audit Certificate

A declaration of confirmation by the auditor that an audit has taken place [3].

C. Audit Report

A written evaluation by the auditor of the results of the audit [3].

D. Audit Trail

Documentation that allows reconstruction of the course of events [3].

E. Clinical Trial/Study

Any investigation in human subjects intended to discover or verify clinical, pharmacological, and/or other pharmacodynamic effects of an investigational product(s), and/or to identify any adverse reactions to an investigational product(s), and/or to study absorption, distribution, metabolism, and excretion of an investigational product(s) with the objective of ascertaining its safety and/or efficacy. The terms "clinical trial" and "clinical study" are synonymous [3].

F. Contract Research Organization

A person or organization (commercial, academic, or other) contracted by the sponsor to perform one or more of the sponsor's trial-related duties and functions [3].

G. Essential Documents

Documents that individually and collectively permit evaluation of the conduct of a study and the quality of data produced [3].

H. Electronic Record

Any combination of text, graphics, data, audio, pictorial, or other information representation in digital form that is created, modified, maintained, archived, retrieved, or distributed by a computer system [CFR part 11.3 (b) [6]] [10].

I. Establishment Inspection Reports (EIRs)

Detailed report of FDA inspection produced after FDA center review.

J. Form FDA 483

Documented inspection observations produced during a FDA inspection.

K. Good Clinical Practice (GCP)

A standard for design, conduct, performance, monitoring, auditing, recording, analyses, and reporting of clinical trials that provides assurance that the data and reported results are credible and accurate, and that the rights, integrity, and confidentiality of trial subjects are protected [3].

L. Good Laboratory Practices (GLP)

Regulations of the US FDA and other countries (based on OECD, Principles of Good Laboratory Practice, 1997) that spell out the requirements for nonclinical (animal or laboratory) studies that will be submitted to the regulatory agency to support a marketing application. US GLPs are found in 21 CFR Parts 58.M [1,2].

M. Good Manufacturing Practices (GMP)

U.S. regulations in 21 CFR Parts 210 and 211 contain the minimum current good manufacturing practices for methods, facilities, and controls to be used for the manufacture, processing, packing, or holding of a drug to assure that it meets the requirements of the Federal Food, Drug, and Cosmetics Act for safety and has the identity and strength and meets the quality and purity characteristics that it purports or is represented to possess.

N. GxP

The acronym GxP represents any combination of the GCP, GLP, and GMP regulations.

O. Inspection

The act by a regulatory authority(ies) of conducting an official review of documents, facilities, records, and any other resources that are deemed by the authority(ies) to be related to the clinical trial and that may be located at the site of the trial, and the sponsor's and/or contract research organizations (CROs) facilities, or at other establishments deemed appropriate by the regulatory authority(ies) [3].

P. Installation Qualification (IQ)

A procedure that describes and documents the necessary steps to install and configure a system. Documented verification that all key aspects of the installation adhere to approved design intentions according to system specifications.

Q. Operational Qualification (OQ)

A procedure that describes and documents the necessary tests done to ensure the installation of a system was successful. Documented verification that each unit or subsystem operates as intended throughout its anticipated operating range.

R. Performance Qualification (PQ)

Documented verification that the integrated system performs as intended in its normal operating environment.

S. Predicate Rule

Existing regulations and/or laws as defined in the Federal Food, Drug, and Cosmetic Act and public Health Service Act (eh 21 CFR 58, 211 and 820). Any preexisting FDA rule that must be complied with to develop, test, manufacture, or distribute an FDA-regulated article.

T. Quality Assurance (QA)

All those planned and systematic actions that are established to ensure that the trial is performed and the data are generated, documented (recorded), and reported in compliance with GCP and the applicable regulatory requirement(s) [3].

U. Quality Control (QC)

The operational techniques and activities undertaken within the QA system to verify that the requirements for quality of the trial-related activities have been fulfilled [3].

V. Regulatory Authorities

Bodies that have the power to regulate. In the ICH GCP guideline, the phrase "regulatory authorities" includes the authorities that review submitted clinical data and those that conduct inspections. These bodies are sometimes referred to as competent authorities [3].

W. Software

A collection of programs, routines, and subroutines that control the operation of a computer or a computerized system.

X. System Development Life Cycle (SDLC)

The period of time that starts when a software product is conceived and ends when the product is no longer available for use or is retired. The SDLC typically includes a concept phase, requirements phase, design phase, implementation phase, test phase, installation, operation, and maintenance phases.

Y. Sponsor

An individual, company, institution, or organization that takes responsibility for the initiation, management, and/or financing of a clinical trial [3].

Z. Standard Operating Procedures (SOPs)

Detailed, written instructions to achieve uniformity of the performance of a specific function [3].

AA. Validation

Establishing documented evidence that provides a high degree of assurance that a specific system will consistently produce a product meeting its predetermined specifications and quality attributes.

BB. Warning Letter

A warning letter is a written communication from the FDA notifying an individual or firm that the agency considers one or more products, practices, processes, or other activities to be in violation of the Federal FD&C Act, or other acts, and that failure of the responsible party to take appropriate and prompt action to correct and prevent any future repeat of the violation may result in administrative and/or regulatory enforcement action without further notice [13].

REFERENCES

1. OECD, Organisation for Economic Co-operation and Development. Principles of Good Laboratory Practice (as revised in 1997). Available at http://www.oecd.org.
2. FDA. 21 CFR Part 58 Good Laboratory Practice Regulations: Final Rule; Department of Health and Human Services: Federal Register: Washington, DC, Sept 4, 1987.
3. ICH International Conference on Harmonization/European Federation of Pharmaceutical Industries Associations. ICH E6 "Good Clinical Practice: Consolidated Guideline." ICH secretariat c/o IFPMA, Geneva, Switzerland, 1996.
4. FDA. 21 CFR Parts 50, 54, 56 and 312: New Drug Product Regulations, Final Rule; Department of Health and Human Services: Federal Register: Washington, DC, March 19, 1987.
5. CAP, College of American Pathologists. Laboratory Accreditation Program, Laboratory Checklists. Northfield, IL, 2000. Available at http://www.cap.org/htm/ftpdirectory/checklistftp.html.
6. ISO (International Organization for Standardization)/IEC (International Electrotechnical Commission). 17025:1999(E). *General Requirement for the Competence of Testing and Calibration Laboratories*; ISO: Geneva, Switzerland, 1999.
7. FDA. Bioresearch Monitoring Program (BIMO), Compliance Program 7348.810, Sponsors, Contract Research Organizations and Monitors. Department of Health and Human Services: Washington, DC, Feb 21, 2001.
8. FDA. Guide to Inspection of Computerized Systems in Drug Processing. US Government Printing Office: Washington, DC, 1983-381-166.2001, Feb 1983.
9. FDA. Technical Reference on Software Development Activities; Reference Materials and Training Aids for Investigators. US Government Printing Office: Washington, DC, July 1987.
10. FDA. 21 CFR Part 11: Electronic Records: Electronic Signatures: Final Rule; Department of Health and Human Services: Federal Register: Washington DC, March 20, 1997.
11. FDA. Guidance for Industry: Computerized Systems Used in Clinical Trials; US Department of Health and Human Services: Washington, DC, April 1999.
12. FDA. Compliance Policy Guide, Enforcement Policy: 21 CFR Part 11: Electronic Records: Electronic Signatures (CPG 7153.17); US Department of Health and Human Services: Washington, DC, May 1999.

13. FDA. FDA/ORA Regulatory Procedures Manual, Chapter 4: Advisory Actions, Subchapter: Warning Letters; US Department of Health and Human Services: Washington, DC, Aug 1997.

11

Biomarker Strategies in Drug Development: Striking a Balance Between Opportunities and Liabilities

Robert A. Dean
Lilly Research Laboratories and Indiana University School of Medicine, Indianapolis, Indiana, U.S.A.

I. INTRODUCTION

Recent advances in genomics, proteomics, combinatorial chemistry, drug screening, and other technologies have expanded the number of therapeutic targets and molecular entities being pursued by scientists in academic, pharmaceutical/biotechnology, and governmental organizations. The expanded portfolios of candidate therapeutics and the dramatic growth of research and development costs relative to growth in sales compel the pharmaceutical and biotechnology industries to pursue a broad range of strategies to improve their return on investment [1]. Academic and government researchers and regulatory authorities are similarly pressured to contain costs while accelerating the delivery of high-quality public services. Strategic use of biomarkers and surrogate endpoints represents one approach being promoted to improve productivity in the discovery, development, and approval of new therapeutics. The Food and Drug Administration Modernization Act of 1997 (FDAMA '97) provides legislative pressure for approval of a new drug product based on a surrogate endpoint that is reasonably likely, based on epidemiological, therapeutic, pathophysiological, or other evidence, to predict clinical benefit in otherwise adequate and well-controlled trials [2]. Moreover, FDAMA '97 mandates that the Secretary of the Department of health and Human Services shall "establish a program to encourage the development of surrogate endpoints that are reasonably likely to predict clinical benefit for serious or life-threatening conditions for which there exists significant unmet medical needs"[3]. In response to this challenge, the FDA partnered with the National Institutes of Health, academia, and the pharmaceutical and biotechnology industries to identify how this mandate might be addressed [4]. The result is an intensified far-reaching consideration of how biomarkers can support drug development and the approval process. This broader topic has promoted the identification, development, and application of biomarkers to more effectively address changing public demand, clinical needs, business environment, and regulatory accountability [5]. With time, energy, and financial investment, these new partnerships will identify and refine biomarkers to address the legislative mandate of FDAMA '97. This chapter provides background information on biomarker-based research in drug development and includes (1) key definitions and scope, (2) an overview of potential opportunities and liabilities related to the use of biomarkers, and (3) a number of strategies that enable research organizations to maximize the potential and minimize the liabilities associated with use of biomarkers.

II. DEFINITIONS AND SCOPE

Simply put, biomarkers are research tools that detect and/or measure biological conditions and events. The National Institutes of Health Working Group on Definitions has provided the following, more detailed definition: a characteristic that is objectively measured and evaluated as an indicator of normal biological processes, pathogenic processes, or pharmacological responses to a therapeutic intervention [6]. This definition highlights the linkage between mechanistic biological models of disease and treatments and the empirical understanding of results gained through randomized clinical trials. This definition further emphasizes the use of biomarkers as clinical and experimental tools that define disease predisposition, diagnosis, staging, and progression and predict and monitor response to therapeutic intervention. The Biomarker Definitions Working Group also define and distinguish biomarkers from clinical endpoints. The latter is a characteristic or viable that reflects how a patient feels, functions, or survives. The definition of a biomarker provides for a broad array of measures as highlighted in the various chapters of this text. This includes, but is not limited to, measures of physiological parameters (blood pressure, heart rate), behavioral tests (measures of cognitive), electrophysiological recordings (electrocardiogram, electroencephalogram), clinical imaging techniques (x-ray, CT, PET, SPECT), and laboratory tests.

III. OPPORTUNITIES

When appropriately applied, biomarkers provide a means to more quickly and reliably characterize experimental models and candidate drugs. This information serves to better define the probabilities of technical and market success and can improve project prioritization, so that the finite resources of a research organization can be more effectively channeled to support drug development and registration. In this manner, an organized biomarker research program can enhance the overall quality of a candidate drug portfolio [7]. The utility of biomarkers, captured in part by the definition above is more fully characterized by a general classification scheme based on three broad experimental applications [8].

Natural history markers characterize individuals with respect to disease risk, diagnosis, and/or progression. Apolipoprotein E and Factor V Leiden genotype are examples of natural history markers for the risk of Alzheimer's disease and thomboembolic disease, respectively. Histological confirmation of neoplasia is a dignostic biomarker and serum βHCG and α-fetoprotein are markers of neoplastic progression of testicular carcinomas secreting these peptides. In clinical trials, natural history markers serve as enrollment criteria for selecting and enriching study populations. Natural history biomarkers provide greater confidence that a study population is appropriate to the posed hypothesis. For example, hypothyroid patients are excluded from studies of antidepressants and antipsychotics based on serum thyroid-stimulating hormone. Hypothyroidism is not the disease under investigation and return to a euthyroid state might correct the psychiatric disorder. Similarly, creatinine clearance is often used to ensure patients with renal insufficiency are not subject to unacceptable risks. Alternatively, natural history markers provide a basis for the stratification of safety and/or efficacy data. While the examples above are clinically oriented, nonclinical correlates of natural history markers are commonly used to evaluate, control, and monitor experimental models.

Biological activity markers reflect efficacy and safety exposure-responsiveness. These responses are used to establish biological/clinical proof-of-concept for a specific intervention, define pharmacodynamic or toxicodynamic relationships, and optimize dose with respect to efficacy and/or safety. This information provides a more accurate estimate of safety margins and, thus improves the probability of success. Biological activity markers can also reduce the development cycle time by allowing for studies of shorter duration or requiring a smaller number of subjects or patients. The earlier and more broadly these risks and benefits can be realized in drug discovery and development, the greater the impact on project-specific resource commitments, optimized study design, clinical plan and time to market, and the overall enhancement of portfolio management.[7] Markers of biological activity also can impact marketing by reassuring patients and their physicians that specific interventions are active. This is particularly important in the setting of chronic therapies, where the clinical outcome is temporally remote or otherwise difficult to monitor. Examples of drug activity markers include measures of viral load to assess therapeutic intervention in hepatitis or HIV infection, and measures of bone mineral density or biochemical markers of bone turnover following treatment for osteoporosis. Biological activity markers also are useful in setting an individualized dose. Hemoglobin A_{1c} is a commonly employed example used to optimize the dose of insulin for management of diabetes mellitus. Biological activity markers can also be used to assess combined dosing of two or more agents as is common with chemotherapeutics cocktails for treatment of neoplasia. As with natural history markers, there are nonclinical correlates of biological activity markers and the latter typically serve as experimental endpoints.

A *surrogate* is a biological activity marker that substitutes for a clinical endpoint [6]. Surrogate endpoints provide an early prediction of the effect of a specific intervention on experimental or clinical outcome. Only a limited number of purported surrogates are widely accepted as clinical endpoints. Blood pressure is one such widely accepted surrogate. Abnormally high blood pressure is consistently associated with the risk of atherosclerotic cardiovascular disease and stroke. Moreover, correction of hypertension reliably predicts therapeutic benefit in stroke prevention [9,10]. Surrogates, especially those accepted by regulatory authorities, provide opportunities for accelerating drug approval. This is particularly true when clinical effects are temporally remote or detection of a clinical effect requires large, lengthy trials [11]. While great emphasis has focused on measures of efficacy, surrogate markers can support phamacovigilance, providing early and more accurate prediction of adverse events. Whether used for prediction of efficacy or safety, for a surrogate endpoint to be useful, the clinical endpoint, class of intervention, and population must be specified [6]. While each class of biomarker provides specific opportunities to improve drug discovery and development, when broadly and strategically applied, these tools can enhance both portfolio management and regulatory decision making [5].

IV. STRATEGIES FOR REALIZING OPPORTUNITIES AND MINIMIZING LIABILITIES

The use of biomarkers to improve and accelerate drug discovery, development, and approval carries with it a number of potential liabilities. These often manifest as poor correlation to an experimental or clinical endpoint and often reflect analytical inadequacies, excessive biological variability, or unrealistic demands on an analytical tool for particular biological or experimental applications. A significant investment in biomarker research and application is required to consistently avoid these problems and realize the opportunities provided by biomarkers. The failure to balance this investment versus the value in productivity is itself a potentially significant liability. The discussion that follows attempts to characterizes various liabilities tied to the use of biomarkers and describe how these liabilities might be minimized.

An acceptable return on biomarker investment requires clearly defined organizational objectives, focused strategies, and effective implementation to deliver on those objectives. Indeed, the investment in biomarker research must be challenged if the fundamental value cannot be clearly defined [6,12,13]. In attempting to assess and assure value, it is useful to determine, at the outset and intermittently, if a given biomarker or biomarker program is (1) analytically sound, (2) biologically and preferably clinically relevant, (3) experimentally operationally practical, (4) correctly interpretable, (5) capable of defining probabilities of success, driving key decisions, and contributing to productivity, and (6) financially reasonable and adequately funded. There are as many different organizational approaches to these evaluations as there are organizations. Nevertheless, organizational biomarker programs need to address two primary needs. First, biomarker strategies need to advance the development of individual programs. Accountability for program-specific biomarker plans and strategies needs to be well integrated into the overall preclinical and clinical drug development plan. This requires that team-specific plans be reviewed and updated in an iterative fashion to address evolving needs. Second, governance over biomarker discovery, development, and implementation needs to be aligned with the governance of portfolio management. As with other program and governance activities, organizational roles, responsibilities, and accountabilities for planning and implementation must be clearly defined.

The goal of biomarker analytical development and validation is to establish a method that is acceptable for the intended application rather than simply validate a developed method [14,15]. Analytical validation typically is defined in terms of analyte specificity and other quantification parameters. It is important that these parameters be clearly defined. Some have suggested that analytical validation of biomarkers, specifically surrogate biochemical markers, should follow good laboratory practice (GLP) like criteria applied to bioanalytical (drug) methods supporting pharmacokinetics [16]. A GLP-like approach provides much-needed discipline to the validation process. Unfortunately, the cost of comprehensive GLP-like analytical validation for all stages of drug development is prohibitive. Moreover, many aspects of the GLP validation processes designed for bioanalytical assays are incompatible with biomarkers based on behavioral, electrophysiological, imaging, or other parameters and enabling technologies. The same is true even for selected laboratory assay-based biomarkers for which no purified or a consensus reference material exists. Despite these limitations, the approach to analytical validation of biomarkers needs to assure that the method detects or measures the analyte of interest in a meaningful and reproducible manner sufficient for the intended experimental application. Thus, the analytical characteristics and expectations for a given biomarker are shaped by the specific technology, whether the methods are qualitative, semiquantitative, or quantitative. The risks and ramifications of false-positive and false-negative conclusions also influence the analytical requirements of a biomarker method. In early drug discovery employing expression profiling, there typically is little need for quantitatively robust methodologies. At this stage, a fold change in a biological response might simply warrant closer examination. By contrast, a biomarker used to determine significant future investment, define a clinical dose as optimally safe and efficacious, or base regulatory approval requires more rigorous analytical validation. Accordingly, research organizations can benefit from the following: (1) Define general guidelines for analytical validation that are technology, stage, and application appropriate. A detailed consideration of this issue is discussed in Chap. 6. (2) Provide individual drug development teams ready access to the resources and consultative support required for consistent application of validation guidelines. (3) Iteratively review the adequacy of analytical validation in a way that ensures a longitudinal view, a risk/benefit assessment, and supports portfolio management.

Transition of biomarkers across stages of drug discovery, development, and even marketing are often unnecessarily complicated by differences in the experimental species, matrix, and methods. Use of common reagents, analytical platforms, and procedures serves to simplify these transitions, reduces the developmental costs, and often provides better longitudinal consistency of experimental data. Centralized clinical laboratories routinely provide longitudinally consistent data that are combinable across study sites, protocols, and drug development programs (see Chap. 2). While this approach is viewed as common for clinical laboratory safety data, the benefits are equally valuable for novel biomarkers. For example, the flow-based Luminex platform and recently marketed reagents allow for the simultaneous analysis of multiple analytes from a single, small-volume sample. Reagents are available for a range of sample matrices and species. Use of this platform and accompanying reagents greatly facilitates the transitions from in vitro to in vivo experiments and from studies of one species to those of another. Similar progress is being made with nonlaboratory biomarkers, notably electrocardiography and molecular and functional imaging.

Poor predictive value for existing conditions and poor correlation to experimental or clinical endpoints can be due to excessive biological variability or inadequately defined exposure-response. These problems can render experimental application of a biomarker impractical or, worse, misleading. The Cardiac Arrhythmia Suppression Trial (CAST) is a commonly cited example of an inadequately defined response. In this study, a decrease in premature ventricular contractions (PVCs) was falsely assumed to be a surrogate marker, predicting the decrease in the risk of sudden death following acute myocardial infarction. The observed reduction in PVCs in response to encainide and flecainide administration did not predict the change in mortality, which unexpectedly increased [17]. This example highlights the need to characterize the biological behavior of critical biomarkers prior to initiation of key studies. Unfortunately, the level of certainty regarding the predictive correlation between a biomarker and the disease, the impact of treatment, and the experimental/-clinical endpoint can only be based upon documented experimental or clinical evaluations. We should not assume it is safe to use a biomarker for a new application until use for that application has been evaluated [18]. Thus, biomarkers are only as robust as the experimental models in which they are validated. Because "validation" connotes the use of a biomarker to be generalizable to other interventions, the Biomarker Definitions Working Group deemed the term unsuitable and prefer the term "evaluation" [6].

Biological or clinical evaluation is typically stepwise in nature and requires progressive application of a biomarker to assess the natural history of a disease, measure the biological activity of a new molecular entity, and serve as a substitute for a specific outcome. Acceptance of a surrogate marker to disease endpoint is achieved by publication of multiple prospectively conducted epidemiological, pathophysiological, intervention, and other studies demonstrating statistically significant predictive correlation. Acceptance also requires substantial evidence of causal or mechanistic linkage between the marker and the clinical outcome [6,19]. Thus, the clinical evaluation process for a surrogate marker is typically beyond the scope of the clinical trial program of a single research organization; hence, the opportunity for a single organization to innovate in clinical surrogate markers is limited. Natural history and drug activity biomarkers are more amenable to exploration and innovation by individual organizations than are surrogate endpoints. While there is greater opportunity in these areas, there often also exists substantial scientific uncertainty about the experimental acceptability of novel biomarkers or novel applications. Thus, rigorous experimental and clinical evaluation of natural history and biological activity markers is essential. The evaluation of natural history markers can be carried out in placebo studies or the placebo arm of controlled studies comparing inactive and active treatments. These studies define within- and between-patient biological variability and are important in the powering of studies using biomarkers to detect pharmacodynamic activity. For biological activity markers the dose response, dose duration, and time to onset and washout of relevant response should be determined. This information is important in the effort to establish or predict mechanistic proof-of-concept and a dose range providing optimal safety and efficacy in humans.

While each drug development program has team-specific needs, development of novel biomarkers or novel approaches for each team can be prohibitively expensive. One way to ensure a reasonable return on investment is to develop strategies that deliver core competencies broadly supporting investigation of biological processes, specific diseases, or target platforms. This minimizes the potential for lost investment due to termination of individual drug development programs. The required core competencies typically relate to enabling technologies, particularly those supporting investigation of specific diseases, therapeutic targets, and recurring safety concerns. Core competences related to research process and infrastructure also must be accessible. Unfortunately, the resources and expertise required for developing and executing effective strategies are often dispersed across multiple functional areas within an organization. Accordingly, the use of biomarkers in drug development is often tactical, rather than strategic, and segmental organization gives rise to reactive solutions that are typically applied too late or inconsistently in development to be of optimal advantage. To cope with this reality, many research organizations have assigned responsibility for biomarkers to a specific functional component. Other organizations have established multidisciplinary strategy groups to coordinate the cross-functional activities required to develop competencies beyond the scope of any individual project team or functional component. Regardless of the approach, these efforts must avoid duplication of effort, effectively draw upon internal and external expertise, address operational needs and constraints faced by individual functional areas, and better integrate biomarker support across stages of drug discovery and development. A successful program provides institutional memory, efficient dissemination of best research and business practices, and systematically pulls nonclinical research forward for use in clinical studies, and vice versa, to enhance the clinical relevance of nonclinical research.

The utility of specific biomarkers for exploration of normal or pathological processes or pharmacodynamic responses is generally greatest when the marker is mechanistically or otherwise closely tied to the phenomenon under investigation and has been thoroughly evaluated through epidemiological research [11,18]. This is seemingly common fare for diseases with established pathophysiology and complimentary biomarkers. However, complex or other unanticipated physiological, pathological processes or pharmacodynamic responses can confound the interpretation of key endpoints. Moreover, novel therapeutic platforms by definition and many diseases are not well characterized. An increasingly common strategy to address this concern is the development of a cluster of markers, a "toolbox" of sorts, to investigate specific diseases or biological phenomena. The concern that a single or even multiple biomarkers may fail to adequately reflect disease diagnosis, progression, or response to therapeutic intervention has generated growing interest in a variety of profiling strategies. These evolving strategies include genomics [20,21], proteomics [22], metabonomics [23], and a host of other evolving biotechnologies. These approaches provide far broader and more rapid characterization of biological states and responses than has been previously possible. Integration of the data derived from these approaches has enhanced biomarker and therapeutic target discovery and development. In the clinical setting, simultaneous measurement of multiple parameters often improves diagnostic sensitivity and specificity. This is not especially new, but even in clinical medicine the scope of discernible signals and the speed and specificity with which they can be measured are dramatic. One such example is the recently reported "multianalyte" diagnostic providing early detection of ovarian cancer at a stage when therapeutic intervention is more likely to be successful [24]. The more comprehensive characterization of specific diseases and the response to therapeutic intervention is improving approaches to disease diagnosis and patient management.

Assessment of safety is critical in the development of every new molecular entity. While the technologies described above are applicable to safety assessment, some aspects of safety assessment must be tailored to the molecular entity, and the experimental and clinical circumstances. A number of organ-specific toxicities are of recurring concern. Morbidity and mortality due to QT prolongation, hepatotoxicity, renal toxicity, and immunogenicity are noteworthy examples. Development of biomarkers and supporting strategies to detect, verify, monitor, and evaluate commonly recurring safety concerns has great value. The recurring need to address these concerns improves the return on the significant investment by regulatory, academic, and industry organizations in applicable safety biomarkers.

Some of the most significant hurdles in successful use of biomarkers are operational. An analytically sound and biologically relevant biomarker is of no value if the application cannot be effectively integrated into an experimental design. Similarly, an inability to refine and efficiently transfer a biomarker from one application to the next is limiting. Methodological studies designed to assess biological or clinical relevance of a marker provide important opportunities to assess operational and experimental practicality. Clinical pharmacologists play a pivotal role in these assessments and the transition of preclinical biomarkers and methods to the clinical settings [7]. The flexible and exploratory nature of study designs common to phase I trials enable the clinical pharmacologist to apply and evaluate novel biomarkers and novel methods. Many research organizations have established "experimental or exploratory medicine" groups to support this activity. The emphasis on "experimental" or "exploratory" medicine rather than on clinical pharmacology, reflects the increasingly multidisciplinary demands of transitions to humans. These new organizations also reflect the need to address similar transitions at other stages of drug development.

Successful implementation of a biomarker strategy often depends on timely access to expertise that is not core to the competences of the research organization or other resources that are not internally available. Thus, the speed and effectiveness with which an organization can draw upon the expertise, services, or other infrastructure of external consultants, partners, and vendors is important. A detailed discussion of this issue is beyond the scope of this chapter. However, the chapter by Kapke and Dean in this text highlights how centralized laboratories and research service organizations can and routinely do provide this support. Similarly, a detailed consideration of disease-focused collaborations involving government, academic, pharmaceutical, and biotechnology organizations is presented in the chapter by Downing.

The cost of methodological and/or biological studies to evaluate efficacy and safety biomarkers can be significant. Ready access to biological specimen repositories from well-characterized disease or other relevant populations can reduce the need for repeated studies of those populations to evaluate the biological/clinical relevance of selected biomarkers. Similarly, historical, population-specific data can provide useful information in evaluating the potential utility of non-laboratory-based biomarkers.

Unfortunately, ready access to this type of biomarker data generated in studies of well-characterized populations is not readily available. Cooperative efforts to create specimen repositories, data registries, and clinical research networks available to multiple research organizations are actively being pursued [5]. In the case of biomarker datasets, ready access often is limited owing to a lack of common data standards and structure, even within individual research programs. The use of data standards can minimize the cost and burden of study-specific programming.

Finally, timely and correct interpretation of biomarker-based datasets requires ready access to a spectrum of bioinformatic, statistical, modeling, and simulation tools, expertise in the use of those tools, and a thorough understanding of the relevant biology and clinical medicine. The initial discovery and evaluation of biomarkers from massive datasets derived from new profiling technologies such as genomics and proteomics are increasingly dependent upon advances in bioinformatics. Subsequent development and evaluation of markers has been promoted by modeling and simulation based on clearly defined assumptions that can be tested through iterative experimentation [25]. As noted earlier, many factors can impact biological processes, disease states, and response to intervention. The need to understand and efficiently explore these biological complexities has given rise to in silico modeling. This approach is being employed to capture, mathematically characterize, and iteratively refine the understanding of specific diseases and therapeutic responses. Early availability and consistent application of these tools allow for more robust and longitudinally comprehensive models [26].

The use of poorly characterized biomarkers increases the likelihood of discordant observations. While decision making is typically based upon the preponderance of data, the need to resolve conflicting data can prove difficult and may undermine the credibility of an otherwise sound study. Accordingly, a priori justification for the selected biomarkers, interpretive scenarios, and how data will drive key decisions is important. Thus, the appropriateness of novel markers or novel applications of conventional markers needs to be defined. An iterative approach that consistently addresses analytical, biological, operational, interpretive, and financial issues is useful. A willingness to drive key decisions, particularly decisions to terminate development of a new molecular entity, provides a means to assess confidence in a biomarker method. When refined, validated, and evaluated to this level, biomarkers can contribute significantly to assessments of the probability of success, support decision analysis, improve the design of large, late-phase clinical trials, and enhance productivity. Failure to strike this balance only adds to the significant financial burden of drug development and approval.

REFERENCES

1. *Pharma 2005: An industrial revolution in R&D*; PriceWaterhouseCoopers, 1998.
2. Food and Drug Administration Modernization Act of 1997. Code of Federal Regulaitons, Title 21, Vol. 5, Part 314, Subpart H, Section 510; US Government Printing Office: Washington DC.
3. Food and Drug Administration modernization Act of 1997. Title I, Subtitle B, Section 112; US Government Printing Office: Washington, DC.
4. Downing, G.J. Biomarkers and surrogate endpoints: clinical research and applications. *Proceedings of the NIH–FDA Conference held on April 15–16 1999 in Bethesda, Maryland, USA*; Elsevier: New York, 2000.
5. Varmus, H. Tools and technologies to enhance clinical research: public–private sector collaborations. In *Biomarkers and Surrogate Endpoints: Clinical Research and Applications*; Downing, G.J., Ed.; Elsevier: New York, 2000; xi–xiv.
6. Biomarker Definitions Working Group; Biomarkers and surrogate endpoints: preferred definitions and conceptual framework. Clin. Pharmacol. Ther. **2001**, *69*, 89–95.
7. Rolan, P. The contribution of clinical pharmacology surrogates and models to drug development—a critical appraisal. Br. J. Clin. Pharmacol. **1997**, *44*, 215–225.
8. Mildvan, D.; Landay, A.; De Gruttola, V.; Machado, S.G.; Kagan, J. An approach to the validation of biomarkers for use in AIDS clinical trials. Clin. Infect. Dis. **1997**, *24*, 764–774.
9. SHEP Cooperative Research Group. Prevention of stroke by antihypertensive drug treatment in older persons with isolated systolic hypertension: final results of the Systlic hypertension in the Elderly Program (SHEP). J. Am. Med. Assoc. **1991**, *265*, 3255–3264.
10. Temple, R. Are surrogate markers adequate to assess cardiovascular disease drugs? J. Am. Med. Assoc. **1999**, *282*, 790–795.

11. Rolan, P. Contribution of pharmacology markers. In *Biomarkers and Surrogate Endpoints: Clinical Research and Applications*; Downing, G.J., Ed.; Elsevier: New York, 2000; 27–35.

12. Blue, J.W.; Colburn, W.A. Efficacy measures: surrogates or clinical outcomes. J. Clin. Pharmacol. **1996**, *36* (9), 767–770.

13. Lesko, L.J.; Atkinson, A.J. Use of biomarkers and surrogate endpoints in drug development and regulatory decision making: criteria, validation, strategies. Annu. Rev. Pharmacol. Toxicol. **2001**, *41*, 347–366.

14. Findlay, J.W.; Smith, W.C.; Lee, J.W.; Nordblom, G.D.; Das, I.; DeSilva, B.S.; Khan, M.N.; Bowsher, R.R. Validation of immunoassays for bioanalysis: a pharmaceutical industry perspective. J. Pharm. Biomed. Anal. **2000**, *21*, 1249–1273.

15. Smith, W.C.; Sittampalam, G.S. Conceptual and statistical issues in the validation of analytic dilution assays for pharmaceutical applications. J. Biopharm. Stat. **1998**, *8*, 509–532.

16. Lee, J.W.; Hulse, J.D.; Colburn, W.A. Surrogate biochemical markers: precise measurement for strategic drug and biologics development. J. Clin. Pharmacol. **1995**, *35*, 464–470.

17. Echt, D.S.; Liebson, P.R.; Mitchell, L.B.; Peters, R.W.; Obias-Manno, D.; Barker, A.H.; Arensberg, D.; Baker, A.; Friedman, L.; Greene, H.L. Mortality and morbidity of patients receiving encainide, flecainide or placebo: the cardiac arrhythmia suppression trial. N. Engl. J. Med. **1991**, *324*, 781–788.

18. Kagan, J.M. Biomarkers for HIV/AIDS: challenges ahead. In *Biomarkers and Surrogate Endpoints: Clinical Research and Applications*; Downing, G.J., Ed.; Elsevier: New York, 2000; 18–22.

19. Fleming, T.R.; DeMets, D.L. Surrogate endpoints in clinical trials: are we being misled? Ann. Intern. Med. **1996**, *125*, 605–613.

20. Debouck, C.; Goodfellow, P.N. DNA microarrays in drug discovery and development. Nat. Genet. **1999**, *21*, 48–50.

21. Hockett, R.D.; Kirkwood, S.C.; Mitlak, B.H.; Dere, W.H. Pharmacogenomics in endocrinology. J. Clin. Endocrinol. Metab. **2002**, *87*, 2495–2499.

22. Mitchell, P. A perspective on protein microarrays. Nat. Biotechnol. **2002**, *20*, 225–229.

23. Nicholson, J.K.; Lindon, J.C.; Holmes, E. "Metabonomics": understanding the metabolic response of living systems to pathophysiological stimuli via multivariate statistical analysis of biological NMR spectroscopic data. Xenobiotica **1999**, *29* (11), 1181–1189.

24. Petricoin, E.F.; Ardekani, A.M.; Hitt, B.A.; Levine, P.J.; Fusaro, V.A.; Steinberg, S.M.; Mills, G.B.; Simone, C.; Fishman, D.A.; Kohn, E.C.; Liotta, L.A. Use of proteomic patterns in serum to identify ovarian cancer. Lancet **2002**, *359* (9306), 572–577.

25. Sheiner, L.B.; Steimer, J-L. Pharmacokinetic/pharmacodynamic modeling in drug development. Annu. Rev. Pharmacol. Toxicol. **2000**, *40*, 67–95.

12
Partnerships in Biomarker Research

Gregory J. Downing
National Institutes of Health, Bethesda, Maryland, U.S.A.

I. INTRODUCTION

Discovery of biomarkers* and the evaluation of their reliability for assessing candidate therapies is a multifaceted process that integrates many disciplines, technologies, strategies, and resources. Successful use of biomarkers in the development of therapies for HIV/AIDS, cancer, cardiovascular disease, and degenerative bone disease has heightened interest for their use in research. The potential use of biomarkers as surrogate endpoints[†] in clinical trials has also been recognized. Recently, efforts have been made to coordinate and streamline biomarker research through collaborations and partnerships among a variety of biomedical research and development (R&D) organizations, including academic biomedical research institutions, commercial R&D organizations (pharmaceutical and biotechnology companies), federal government biomedical research and regulatory agencies, and voluntary, not-for-profit health advocacy foundations. The rationale for encouraging partnerships in biomarker research includes the complex and rapidly changing clinical research environment, the high level of complexity of the diseases investigated, a substantial increase in the number of candidate therapies to be evaluated, finite fiscal resources to

---

*Biomarker (biological marker) is defined as a characteristic that is objectively measured and evaluated as an indicator of normal biological processes, pathogenic processes, or pharmacological responses to a therapeutic intervention [1,2].

[†]Surrogate endpoint is defined as a biomarker that is intended to substitute for a clinical endpoint. A surrogate endpoint is expected to predict clinical benefit (or harm or lack of benefit or harm) based on epidemiological, therapeutic, pathophysiological, or other scientific evidence [1,2].

support rising costs of therapeutic R&D, and the lack of high-precision clinical technologies to assess novel therapies in clinical trials.

This chapter explores the interplay among the components of the biomedical research enterprise that underpin the biomarker research. A cornerstone of this discussion is a focus on new partnership models to support biomarker R&D. General concepts of public–private partnerships, including incentives and challenges for these collaborations, are described. Also examined are common questions regarding public–private partnerships in biomarker research such as: What are the opportunities, incentives, and motivations that foster collaborations in biomarker research? Will partnerships in biomarker research spur therapeutic product development, particularly for diseases that have few or ineffective disease-modifying therapies?

II.   AN OVERVIEW OF APPROACHES TO BIOMARKER DISCOVERY AND EVALUATION

Clinical knowledge about biomarkers is derived from data generated in studies that assess biomarker response over the course of disease (natural history studies) or its response to a therapeutic agent (clinical trials). The evaluation of the value of a biomarker to predict clinical response depends on accrual of data, if sufficient data exist to link the biomarker to clinical outcome [3–5]. Rarely is there a single definitive research study that fully defines the association of a particular biomarker to a disease process, and inferences about biomarker application in clinical settings are determined by expert consensus. A less common approach is to plan biomarker development and evaluation strategically, although some corporate R&D organizations now plan biomarker research in tandem with product development. An example of this approach is the development of biomarkers for HIV/AIDS during the 1980s. Given the health emergency at hand, researchers from the public and private sectors used biospecimens in a retrospective fashion to establish the validity of CD4+ cell counts and HIV viral load measurements as guideposts for early efficacy assessment, and later as guideposts for therapeutic management [6–9]. Another commonly cited example of the use of biomarkers in elucidating disease processes involves the role of cholesterol measurements in coronary heart disease provided from large epidemiological studies such as the Framingham heart study and early-phase clinical studies with HMG-CoA reductase inhibitors.[10] Both models exemplify successful approaches to the use of public and private resources in developing useful clinical measures of disease and health states.

Three communities interface in biomarker research: academia, industry, and the federal government (Fig. 1) [9]. In this model, each sector contributes key components that are complementary but not duplicative. The most difficult

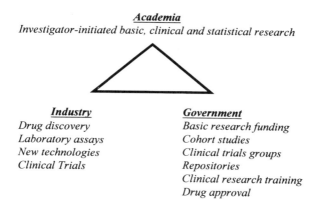

Academia
Investigator-initiated basic, clinical and statistical research

Industry
Drug discovery
Laboratory assays
New technologies
Clinical Trials

Government
Basic research funding
Cohort studies
Clinical trials groups
Repositories
Clinical research training
Drug approval

Figure 1   Biomarker research interactions among the public and private sectors.

challenge with this model of distributed roles is to establish meaningful utility from the information that emerges from each sector. For example, the federal government may not conduct many single product clinical trials that reveal information about a biomarker. On the other hand, industry is unlikely to focus its resources on longitudinal population studies that provide data about disease progress and clinical outcomes. These data are important in evaluating biomarkers as valid surrogate endpoints. Another aspect of biomarker research in therapeutic development is the interface with regulatory authorities such as the U.S. Food and Drug Administration (FDA). Thus, in the interest of not only the involved parties but also better patient care, the three groups must interface effectively to use biomarkers to enhance the pace of developing therapies. Typically, biospecimens and clinical data are obtained in clinical trials, and natural history studies are conducted by both government and private research organizations. Often, the data are used collectively to make inferences about the conditions in which a biomarker accurately represents a disease or health state and how it is affected by a therapy. Therefore, a combination of resources is used to develop a knowledge base about a biomarker and its utility to predict drug response.

III.   INCENTIVES TO ESTABLISH PUBLIC–PRIVATE PARTNERSHIPS IN BIOMARKER RESEARCH

A highly successful track record for innovative technology development has been established in university laboratories and clinics and corporate R&D companies without formal collaborations. At the same time, federally sponsored basic biomedical research has been lauded for its role as an engine for generating knowledge and research tools for private-sector development of devices, pharmaceuticals, and biological agents. Given this success, what compelling considerations are there to support creation of public–private partnerships as a means of providing a strategic advantage in biomarker research?

There are a variety of reasons. One is that biomarker research crosses many planes of the research enterprise. Many investigators are supported by sponsored research agreements from private industry as well as by federal grant and contract awards. Moreover, there is a paucity of precise clinical measures for many pathogenetic components of disease pathways despite advances in understanding of cellular molecular, and genetic factors. Among the research areas that are frequently mentioned in need of biomarkers are aging, neurodegenerative disorders, inflammation, cancer, cardiovascular disorders (e.g., congestive heart failure, atherosclerosis, peripheral vascular disease), autoimmune disorders, and mental health disorders. Basic laboratory tools have been proposed as an avenue to monitor clinical disease. Such tools include large-scale protein separation and identification techniques, DNA microarray technologies, high-resolution-imaging instruments such as positron emission tomography, magnetic resonance, mass spectrometry, and cellular imaging techniques. However, the customization of these tools for clinical use and the standardization of their measurements are crucial factors for their successful use. Early-phase clinical trials are often good avenues for clinical specimen and data collection, but often these are performed in small, homogeneous populations that are not conducive to fully evaluating a biomarker.

Development of clinical measurement tools has not held a high priority for federally sponsored biomedical research. Biomedical research funded by the National Institutes of Health (NIH) is supported by a variety of mechanisms, and research project grants (RPGs) are the most common funding mechanism. Investigator-initiated research is typically supported through grants to institutions (universities, medical centers, and other research organizations). The major portion of these proposals is a hypothesis-based, competitive, and highly innovative research plan. The transfer of laboratory technologies from basic research laboratories to clinical research is commonly viewed as applied research. Therefore, research on the discovery and evaluation of clinical biomarkers is rarely a candidate for this type of funding. On the other hand, federal resources do support small-business innovation research (SBIR) grants and small-business technology transfer (STTR) awards. Other mechanisms that the government uses to support technology development in the private sector include cooperative research and development awards (CRADAs), and contracts that are often used to develop clinical assays and analytical tools.

Another incentive to form partnerships in biomarker research is to compensate for the changes in the health care delivery system that have weakened the nation's clinical research infrastructure. The introduction of managed care in many commercial markets in the 1980s and 1990s dampened clinical research at many academic medical centers. In some cases, managed care has yielded disincentives for health care providers and patients to engage in clinical protocols, resulting in a smaller participant pool for clinical research. Another indirect effect of managed care on clinical research is the increased demands placed on academic physicians for clinical service delivery. This has led to fewer skilled clinical investigators with less dedicated research time concurrent with an increased need for such investigators.

Over the past several years, discussions with scientists from industry, contract research organizations, academia, and government research and federal regulators have explored options for collaboration. These discussions provided ample reasons to explore the organization of research partnerships, including the following:

Foster innovative approaches to clinical measurement

Develop strategies to improve efficiency and safety of therapeutic development by implementing biomarkers in preclinical and clinical studies

Encourage sharing of research resources (e.g., biospecimens, databases, imaging repositories, etc.)

Establish an infrastructure of publicly available databases and specimen repositories to support biomarker research without impediments to technology licensing rights

Enhance scientific communications about biomarker research

Catalyze research interest in "underdeveloped" disease areas with few or no existing effective therapies

Facilitate validation methods and clinical standards development

Leverage fiscal resources

Support reference datasets for clinical trials and supporting data for establishing surrogate endpoints

Develop infrastructure that will support development and assessment of new technology platforms

IV. NEW APPROACHES TO BIOMARKER RESEARCH: PUBLIC–PRIVATE PARTNERSHIPS

Recognizing the large-scale needs and costs, requirements for clinical and technological expertise, and importance in streamlining development, leaders in the biomedical R&D enterprise initiated a series of discussions in May 1997. These discussions were coordinated by then Director of the National Institutes of Health (NIH), Harold Varmus, in an effort to revitalize clinical research and streamline therapeutic R&D. Included in these discussions were chief executive officers of major pharmaceutical and biotechnology companies, the Commissioner of the U.S. Food and Drug Administration, representatives from contract research organizations, academic medical centers, third-party payers, academic researchers, and others. By examining common interfaces in research, the group identified opportunities to work collaboratively to streamline therapeutic development and testing. Eight areas were identified as possible targets for collaboration, and among them was a research focus on biomarkers and surrogate endpoints. A variety of planning meetings and conferences were convened, and papers were then prepared to focus attention on the need for clinical assessment tools. Key disease areas were identified where biomarker research would be of high value to therapeutic developers, and high-priority areas included cancer, inflammatory and autoimmune disorders, neurodegenerative diseases, bone and joint degenerative diseases, bone and joint degenerative diseases, mental health disorders, and diseases associated with growth and aging. Over the course of the past 3 years, the NIH released many requests for proposals and program announcements encouraging research in these areas. Many corporate entities focused attention on biomarkers and began integrating them into their R&D and clinical trials strategies.

What evidence exists to suggest that collaborations among public and private research organizations in biomarker research will succeed? On a broad scale, public support of basic research and collaboration with private companies in the United States led to the development of the military-industry enterprise in the 1940s. Historians point to the 1945 report "Science—The Endless Frontier," prepared by Vannevar Bush [11], then Director of the Office of Scientific Research and Development, as the seminal event leading to the public policy of government-funded research at universities to increase technological development by industry. This policy guided national interest in the public support of biomedical research that is now considered a vital interest of the federal government.

Regarding public–private partnerships in biomedical research, there are several recent examples. One intriguing aspect of such partnerships is that competitors from the private industry sector become collaborators. Clearly, an incentive for attracting private interests to work collaboratively is the nature of the research that the project will encompass (i.e., consistent with a company's corporate R&D plans). Research suited for partnerships has qualities of being "precompetitive," is not oriented toward development of a specific product or clinical trial or a particular new drug entity, and creates a research infrastructure by not placing any one company at a competitive advantage over another. An example of such a partnership is the Single Nucleotide Polymorphism (SNP) Consortium,

Ltd., a nonprofit foundation organized to create and make publicly available a high-quality SNP map of the human genome. SNPs may be shared among groups of people with harmful, but unknown, mutations and serve as markers for them, and identification of such markers may help identify the mutations and accelerate efforts to find therapeutic drugs. The consortium's members include the medical research charity the Wellcome Trust and 12 pharmaceutical and technology companies: Amersham Pharmacia Biotech, AstraZeneca PLC, Aventis Pharma, Bayer AG, Bristol-Myers Squibb Company, Glaxo Wellcome, F. Hoffman-La Roche, IBM, Novartis, Pfizer Inc., Searle, and SmithKline Beecham PLC. Academic centers, including the Whitehead Institute for Biomedical Research, Washington University School of medicine in St. Louis, the Wellcome Trust's Sanger Centre, Stanford Human Genome Center, and Cold Spring Harbor Laboratory, are also involved in SNP identification and analysis. Both the protocols and the database of SNPs are freely available to scientists.

Another example of public–private partnership in basic research is the Mouse Sequencing Consortium, which was established to decode the mouse genome. Sponsors of this effort included SmithKline Beecham, the Merck Genome Research Institute, the Wellcome Trust, and the NIH. Collectively, these organizations developed a new model for large-scale genomics research that yielded a genomic map for a strain of mice commonly used in biomedical research.

Government and industry partnerships are also developing on a global scale to meet pressing public health needs. The current landscape in international health shows a proliferation of global public–private partnerships created to address numerous issues from expanded access to drugs and vaccines in poor countries [12,13] to the highly publicized Bill and Melinda Gates Foundation's relationship with industry to develop AIDS vaccines (www.gatesfoundation.org) [14]. Other international initiatives are underway for collaborations with governments and industry in developing therapies for malaria and tuberculosis. Moreover, World Health Organization (WHO) president Gro Harlem Bruntland has suggested recently that the organization must establish ties with the private sector and industry because "the broad health agenda is too big for the WHO alone" [15]. These partnerships hold great promise as possible solutions to long-standing problems [16,17].

V. PRINCIPLES OF PUBLIC–PRIVATE PARTNERSHIPS IN BIOMEDICAL RESEARCH

Recent discussions regarding the establishment of biomarker research partnerships have provided insights into the scientific issues and administrative policies in uniting organizations to accomplish a common goal. Scientists from academia, industry, and government recognized that there are different motivating factors for their participations and that it was important for them to understand core issues of partnership. However, some common principles are emerging that underpin successful partnerships. In fact, Reich [18] suggests that a working definition of public–private partnership must contain three core points. First, partnerships involve at least one private for-profit organization and at least one not-for-profit organization. Second, core partners share effort and benefits jointly. Third, partnerships centered on biomedical research should have a socially based goal related to the improvement of some aspect of health, especially for disadvantaged populations. Organizations involved in public–private partnerships may have different motivating factors for participating in the venture. Thus, an effective partnership between diverse organizations must be carefully constructed and often requires restructuring as other groups become involved.

Partnerships confront seven organizational challenges that Austin [19] calls "the seven c's of strategic collaboration": clarity of purpose, congruency of mission, strategy, and values, creation of value, connection with purpose and people, communication between partners, continual learning, and commitment to the partnership. Reich [18] notes that the challenge of creating value is of special importance in public health partnerships, for the value created must be useful not only to core partners but to an entire society as well. Thus, public–private partnerships are learning processes and should be seen as fluid, evolving entities rather than fixed agreements.

VI. GOVERNANCE OF PARTNERSHIPS

Successful partnerships operate with principles for governance to clarify the roles and responsibilities for each party. Governance can be defined as "the process whereby an organization or society steers itself" and is usually assisted by a system of rules, norms, and processes through which power and decision making are exercised. Good governance is considered to have four components: (1) representative legitimacy; (2) accountability; (3) competence and appropriateness, and (4) respect for due process. Legitimate representation addresses the issues of whose interests should be represented in the partnership and whose should not be. Each party involved in a partnership should have representation in the organizational structure. Representation also entails accountability that on a broad level refers to holding responsible parties accountable for their actions. In biomarker partnerships, a program manager is usually the key facet for facilitating communications. Competence and appropriateness pertain to the attainment of expert input into the management of the projects. In the case of biomarker partnerships, each party usually brings special expertise on scientific and administrative matters, and consultations with experts in the research community are commonplace. Due process is owed to all parties in public–private partnerships, and this is achieved by transparency and public disclosure about the partnership.

VII. OTHER INCENTIVES FOR BIOMARKER COLLABORATIONS

Several other factors have affected the conduct of clinical research and the opportunities to conduct biomarker research. First, vast changes in the clinical research arena as well as in the pharmaceutical and biotechnology industries have followed these rapid changes in knowledge and technology. For example, the role of contract research organizations in providing infrastructure support to clinical trials rose dramatically during the 1990s. At the same time, academic research institutions experienced a major upheaval in health care delivery through the introduction of nationwide managed care programs [20]. As a result, many clinical research programs were left bereft of faculty or opportunities to conduct clinical research. Unfortunately, this shift occurred at a time when clinical research expertise was in high demand. Furthermore, acquisitions, consolidation, and mergers of major pharmaceutical firms in the 1990s may have created fewer drug development opportunities and reduced the availability of resources needed for R&D of new therapies. On the other hand, this decline was countered by a substantial increase in private-sector R&D spending during this time. According to the Pharmaceutical Research and Manufacturers of America (PhRMA) 2000 Industry Profile, research-based companies have more than tripled their R&D expenditures since 1990 (www.phrma.org/publications/publications/annual2001). Moreover, the number of new chemical entities approved by the FDA has increased substantially in the 1990s as compared to previous decades (www.fda.gov/cder/rdmt/fy00ndap.htm). In summary, there has been a recent surge in therapeutic developments, although the infrastructure to support clinical assessment of them is on uncertain ground.

VIII. CHALLENGES TO PUBLIC/PRIVATE COLLABORATIONS FOR BIOMARKER RESEARCH

When opportunities for public–private collaborations are examined, several major issues emerged as challenges, including intellectual property associated with biomarker development, conflicts of interest, and return on investment. Some of the major policy issues are listed below, and two major issues, intellectual property and conflict of interest, will be discussed in detail.

Intellectual property and material transfer agreement practices
Equitable sharing of resources
Mechanisms of financial support
Representation in decision making
Conflicts of interest in research
Return on investments (public and private)
Restrictions on private partner participation in government review and
     awards practices
Protection of privacy and confidentiality for human subjects information

Discussed here are two major issues that have been the focus of specific
attention during the course of planning biomarker partnerships.

A.   Intellectual Property Issues and the Bayh-Dole Act

Technology transfer—the transfer of research results and technologies from
universities to the commercial sector—has been an issue for academics and
lawmakers for more than 40 years. Although the Manhattan Project demonstrated
the federal government's potential to share technology, much debate remained in
the decades following World War II regarding federal patent policy, which
eventually resulted in legislative action. Because there was no government-wide
policy regarding ownership of inventions made under federal funding, the
diversity in policies among the various funding agencies greatly hampered the
transfer of government-assisted inventions to the private sector.

The problem was due mostly to the government's reluctance to relinquish
ownership of federally funded inventions to the universities or other grantees that
developed them. Instead, the government made such inventions available by
nonexclusive license to any interested party, and an organization therefore could
not obtain exclusive rights to manufacture and sell a resulting product.
Understandably, companies had little enthusiasm for developing early-stage
inventions; when products finally were ready to reach the market, competitors
could acquire a license and then manufacture and sell the same products. After
several decades of such inefficient policy making, the administration proposed
legislation in 1980 that encouraged the utilization of inventions produced under
federal funding and promoted the participation of universities and small
businesses in development and commercialization processes. Known as the
Bayh-Dole Act (Public Law 96-517), this law provides the basis for current
university technology transfer practices [21].

The Bayh-Dole Act contained several important provisions. First, it
established a uniform federal patent policy that encouraged universities and
nonprofit organizations to collaborate with industry to utilize inventions created
using federal funding. Furthermore, it clearly stated that these groups may elect to

retain title to inventions developed through governmental funding, although the government retains a nonexclusive license to practice the inventions throughout the world. On February 18, 1983, a Presidential Memorandum on "Government Patent Policy" was issued, stating that federal agencies were to extend the statutory terms to for-profit grantees/contractors as well.

The Bayh-Dole statute and subsequent amendments created incentives for the government, universities, industry, and the small-business sector, and herein may lie the reason for its success. Data suggest that the Bayh-Dole Act had an immediate impact on technology transfer. For example, between 1974 and 1984, 84 universities applied for 4105 patents (2944 were subsequently issued); in 1992 alone, 139 universities received 1557 patents [22]. Also during 1974–1984, 1058 licenses were granted by universities; in the period 1989–1990, 1510 licenses were granted [23]. These data suggest that the Bayh-Dole Act has promoted a substantial increase in technology transfer from universities to industry and ultimately to the public.

Why is this important to biomarker research partnerships? It is important when planning collaborations to develop the principles under which the technologies produced from research initiatives are shared between collaborators and the research community. Recently, the NIH developed a policy to encourage the sharing of research resources among investigators entitled "Sharing Biomedical Research Resources" (http://ott.od.nih.gov/NewPages/RTguide_final.html). This policy is designed to hasten R&D by discouraging the unnecessary licensing of research tools that by themselves have limited commercial value. This principles strive to discourage time-consuming and costly administrative practices associated with patenting and material licensing agreements. Public–private partnerships should recognize the value of developing policies overseeing biomarker research that do not close off opportunities to advance science and future product development.

B. Conflict of Interest

The interface between industry, academia, and the government has historically been a fluid one, and there has been a great deal of interest lately in examining this intricate web of interactions with regard to clinical trials research [24,25]. As a result, the medical literature contains numerous articles expressing concern about the role of industrial funding in clinical research [26–29]. Although these articles address generalized concerns, the issues raised may also be applied to research in biomarker development. Owing to the increasingly complex financial interactions between investigators, institutions, and patients, conflict of interest has focused on academic/industry collaborations [25,30,31].

A recent study of 89 of the 100 institutions with the most funding from the NIH demonstrated that most policies on conflict of interest lack specificity about the kinds of relationships with industry that may be permitted [32]. A similar study by Lo et al. [33] analyzing policies addressing conflict of interest at the 10 U.S. medical schools receiving the largest amount of NIH research funding showed that only one school met the authors' suggestion that university-based investigators be prohibited from holding stock, stock options, or decision-making positions in a company that may be reasonably affected by the results of their research. To minimize such conflict of interest issues, Moses and Martin [34] suggest several principles that must be considered when developing a working partnership. First, veracity of basic research results must not be compromised. Moreover, a disinterested party should examine the relationship at the outset and at key developmental points to determine the course of action. Proprietary rights and control of intellectual property must be acknowledged at the outset, and assurances regarding the right to publish must be clear. Finally, financial and nonfinancial incentives should be designed to meet the varying needs of all involved parties, including the institution, the senior investigators, and the junior faculty.

IX. EXAMPLES OF PARTNERSHIPS IN BIOMARKERS RESEARCH

Several common themes are observed in disease areas where formation of partnerships for biomarker development is of particularly high interest. Diseases associated with the aging process are highly represented in this scenario, as progression to primary clinical endpoints by its nature requires long study periods for clinical trials, thereby limiting the number of candidate therapeutic options that can be evaluated. Clinical studies of prevention strategies, particularly those with low rates of occurrence of symptomatology, require high numbers of clinical participants. Comorbidity conditions present another challenge to therapeutic development. Some examples of public–private partnerships for biomarker research that address specific disease states are described here.

A. Osteoarthritis Biomarkers

The goal of the Osteoarthritis Initiative (www.nih/gov/niams/news/oisg/index.htm) is to develop and support clinical research resources that enable discovery, assessment, and validation of biomarkers for osteoarthritis. It is a 7-year collaborative project between five pharmaceutical companies—Merck, Pfizer, Novartis, Pharmacia, and GlaxoSmithKline—the Foundation for the NIH, and several NIH institutes. The mechanisms supporting the research infrastructure are shown in Fig. 2. The Foundation for the NIH, established by Congress in 1996 as a 501(c) [3] organization, serves as a nonprofit administrative interface between the government and private industry sponsors. The purpose of the Foundation is to

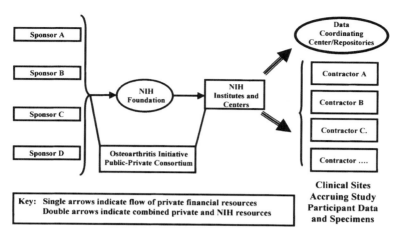

Figure 2 The Osteoarthritis Initiative project model: financial resource management.

foster collaborative relationships in education, research, and related activities between the NIH, industry, academia, and nonprofit organizations.

The osteoarthritis project anticipates development of biomarkers that include biochemical markers of bone and cartilage, genetic markers associated with osteoarthritis, and structural markers determined using various imaging technologies such as radiographs and magnetic resonance. Funded over a 7-year period at an estimated $60 million, the Initiative anticipates that these biomarkers will provide the nonprofit and commercial scientific communities with new opportunities to develop disease-modifying therapies and streamline clinical trials to assess the safety and efficacy of these therapies. In addition, these resources should facilitate the development of in vitro analytical methods useful in patient diagnosis and management.

Numerous benefits are expected, both for the patient population and for the participants in the Initiative. Development and validation of biomarkers will streamline the clinical trial process and provide incentives for private sector R&D of novel osteoarthritis interventions and in vitro diagnostic products. Moreover, private-sector sponsors may collaborate with academic and NIH scientists in the design of the research plan, the use of research tools, and the management of the consortium and the resources developed by it (Fig. 3). Furthermore, participation by regulatory agencies such as the U.S. FDA will improve communication in the regulatory process. As a consequence, patient care is enhanced through streamlined development of regulatory guidelines, enhanced evaluation of

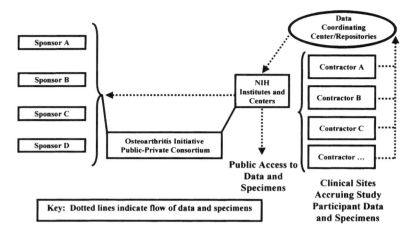

Figure 3   The Osteoarthritis Initiative project model: management of data and specimens.

surrogate endpoints in clinical trials, and a more efficient regulatory approval process for new therapies.

The scientific plan was developed in public meetings and includes the following core goals:

> Develop a population-based, longitudinal, human subject cohort to characterize the natural history of osteoarthritis.
>
> Apply imaging tools (radiographs and magnetic resonance) to evaluate joint structural markers (principally for the knee joints) as potential surrogate endpoints for clinical trials
>
> Establish biospecimen repositories to enable evaluation of biochemical and genetic markers

In this model, a steering group composed of members from public organizations (NIH and academic investigators), private organizations (sponsoring companies), and liaison representation from the FDA will oversee and provide input to the administration of the research activities. The core units of the research network were recently established. A data-coordinating center established at the University of California, San Francisco will assemble the epidemiological data, track biospecimen and imaging data collection, assemble data and disseminate it among the research community, and coordinate research activities across the network. Four clinical sites, at Ohio State University, University of Rhode Island, University of Pittsburgh, and University of Maryland, serve as the major center for the recruitment of the 5000 research participants. Subcontracts have been issued for support services such as maintaining a biospecimen repository and developing quantitative metrics of the clinical parameters measured in the project. Participating members are also providing information to the research community about how the research resources developed through the consortium will be made publicly available.

Participation in the Consortium will allow integration of data and technology into strategic plans for clinical trials, since validation studies will be conducted with the data and results made available in the public domain. This conserves fiscal and human resources by eliminating the need for multiple private companies to repeat assays and redevelop techniques. By establishing the clinical research resources through this consortium, it is anticipated that private industry sponsors will experience a reduction in development costs and time to evaluate new osteoarthritis therapies.

Several key policies emerged from this model. First, databases developed from the research studies will exist in the public domain, and administrative mechanisms will be implemented to preclude the patenting of the public databases. This will enable academic and commercial research interests to use the knowledge and technology to create new intellectual property without being blocked. Second, in this model, biomarker technology and material rights to such technology can be developed under existing patenting and licensing policies. For example, investigators (public or private) who create new assays or measuring technologies through the use of specimens and data developed by the Initiative retain the ability to commercialize their discovery. A requirement for access to specimens and data will be to present a plan to provide equal access to the resource so as to not block further research or commercial use.

B. Cancer Biomarkers

The early detection research network (EDRN) of the National Cancer Institute (NCI) represents a linkage of basic and clinical laboratories seeking to identify biomarkers for various solid tumors. The EDRN is a consortium representing about 30 laboratories across the country, and it is designed to provide a linkage between cancer and biomarker discovery and clinical applications. The management of the consortium is provided by the NCI in consultation with a steering committee composed of the principal investigators from each site. An independent advisory committee (AC) advises the steering committee (SC) and the NCI regarding recent progress in biomarkers research and suggests avenues for the consortium to consider. A data management and coordinating center (DMCC) manages information flow across the centers and laboratories (Fig. 4). Approximately half of the laboratories have sponsored research agreements and technology licensing rights with private companies. Industry scientists are actively engaged with EDRN investigators and exploring scientific collaborations, and they impart knowledge to the network by participating in the organization's scientific meetings and electronic forums.

The EDRN also has an Associate Membership Program that allows interested parties to submit proposals for funding, in particular to develop informatics tools for the consortium. The EDRN currently has two websites (http://cancer.gov/edrn): a password-protected site accessible to EDRN investigators and a public site for information, news, and contacts for the consortium. In addition, the EDRN has been featured in major journals and conducts annual workshops and conferences to investigate new frontiers in cancer detection and diagnosis research.

The EDRN employs a systematic process for taking a biomarker from development to validation that may be well suited for assessing privately developed biomarkers. First, the investigator submits a proposal to the steering committee for review. If it is approved, the biomarkers validation laboratory conducts an assay cross-check.

If the cross-check is approved, study designs and protocols can then be established with the assistance of the SC, AC, and DMCC. The specific goals of the EDRN are as follows:

Identify and validate promising biomarkers for large-scale studies
Conduct early phases of clinical/epidemiological studies

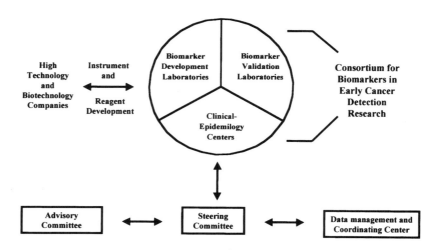

Figure 4   The early detection research network.

Establish an EDRN informatics linkage with the NCI Enterprise System
Formulate a molecular taxonomy of precancerous lesions establishing
standards for precancer classification
Establish standards for analytical and clinical validation of biomarkers

X. OTHER FORMS OF PUBLIC–PRIVATE PARTNERSHIPS SUPPORTING BIOMARKER RESEARCH

The Immune Tolerance Network (ITN) (http://www.immunetolerance.org) is a collaborative network of 40 research institutions that addresses clinical trials in kidney and islet transplantation, clinical trials in autoimmune disease, development of tolerance assays, and clinical trials in asthma and allergic diseases. The ITN is a public–private partnership with funding support provided by two NIH Institutes and the Juvenile Diabetes Research Foundation (JDRF). The network is designed to solicit, develop, implement, and assess clinical strategies and biological assays for the purposes of inducing, maintaining, and monitoring tolerance in humans for these conditions. The network encompasses two different components: (1) clinical trials and mechanistic studies in kidney and islet transplantation, autoimmune disease, and asthma and allergic disease and (2) development and validation of assays to measure the induction, maintenance, and/or loss of immune tolerance in humans. Diseases currently under investigation include type 1 diabetes, rheumatoid arthritis, multiple sclerosis, and systemic lupus erythematosus. Core assay facilities currently include a polymerase chain reaction-based gene expression and polymorphisms core, a pharmacogenomics and microarry core, a mixed histocompatability complex–peptide complex core, and a cell-based tolerance assay core. The ultimate goal is to make as much of the information public as is possible, although the network is still addressing this issue. Administrative mechanisms also exist within the ITN for private-interest collaborations in the research initiatives.

XI. PARTNERSHIPS IN BIOMARKER TECHNOLOGY DEVELOPMENT—THREE MODEL COMPANIES

Three companies, Synarc, Surromed, and Affymetrix, serve as models for successfully interfacing resources from the public and private sectors for the development of biomarkers for clinical trials. Synarc (www.synarc.com), founded in 1998, is a global company designed to bring together medical imaging and molecular marker services to enhance biomarker development and validation and accelerate clinical trials for arthritis and osteoporosis. Synarc provides a complete resource for investigators using imaging markers in a clinical trial by assisting clinicians with all steps of the process, including protocol review and design, patient selection and screening, biomarker assay development and analysis, data management, and presentation of results. Partnered with academia (Stanford University) as well as private-sector investors, Synarc currently manages over 100 clinical trials globally and provides a model for combining scientific, clinical, and data managerial experts to expedite drug development.

Surromed (www.surromed.com), a privately held biotechnology company, develops tools and technologies for comprehensive phenotypic analysis. The company has recently announced a multiyear collaboration with the nonprofit Palo Alto Medical Foundation, a multispecialty group practice with over 400 physicians. The goal of the collaboration is to identify biological markers by conducting clinical phenotyping research in multiple disease areas, beginning with rheumatoid arthritis. The collaboration will combine Surromed's integrated phenotyping platform with the clinical expertise of the Palo Alto Medical Foundation to improve the discovery and development of diagnostic and therapeutic products.

Another unique partnership has been drawn between Affymetrix, Inc. (www.affymetrix.com), a producer of microarray technology, and the Cystic Fibrosis Foundation Therapeutics, Inc. (CFFTI). Under this arrangement, CFFTI offers subsidies for the purchase of specific genome arrays from Affymetrix by academic, not-for-profit research laboratories and discounts for their purchase by certain research units of for-profit entities that have a preexisting relationship with the Cystic Fibrosis Foundation for the purpose of conducting research related to cystic fibrosis. This arrangement will enable researchers to expand their research capacity in identifying and evaluating genes as biomarkers for cystic fibrosis. Affymetrix has agreed to sell arrays to research laboratories and for-profit laboratories selected by CFFTI prior to the general commercial availability of the arrays and at substantially discounted prices. In exchange for receiving the subsidy or the discount, recipients will conduct cystic-fibrosis-related experiments that utilize the arrays and submit data from those experiments to a database managed by CFFTI and the University of North Carolina, Chapel Hill. Each qualified user will have access to this database for the purpose of data mining.

XII. OTHER FEDERAL FUNDING MODELS FOR SUPPORTING PRIVATE-INDUSTRY BIOMARKER RESEARCH

Several mechanisms exist for biotechnology and instrumentation companies to participate in collaborative research with federal government. These include the Small Business Innovation Research Program (SBIR), Small Business Technology Transfer Program (STTR), and Cooperative Research and

Development Agreements (CRADA). Each of these programs has successfully supported development of clinical measuring technologies in the private sector.

A. Small Business Innovation Research Program

The SBIR program (http://grants.nih.gov/grants/funding/sbir.htm) is a set-aside program designed to support innovative, commercially viable research conducted by small business concerns. Innovation and the potential for commercialization are among the important factors included in the review criteria for evaluation. Small businesses in any biomedical or behavioral research area may submit grant applications to the NIH for consideration. Support under the SBIR program is normally provided for 6 months/$100,000 for phase I and 2 years/$750,000 for phase II. In fiscal year (FY) 2000, the NIH made SBIR grant and contract awards totaling $352 million. The amount available for the NIH SBIR program in FY 2001 was estimated to be $410 million.

B. Small Business Technology Transfer Program

The STTR program (http://grants.nih.gov/grants/funding/sbir.htm) is designed to support innovative, commercially viable research conducted cooperatively by a small business concern and a research institution. As in the SBIR program, innovation and the potential for commercialization are two of the review criteria in the evaluation process. However, at least 40% of the research project must be conducted by the small-business concern, and at least 30% of the work must be conducted by the single, "partnering" research institution. The NIH welcomes grant applications from small businesses in any biomedical or behavioral research area. Support for the STTR awardees is normally provided at 1 year/$100,000 for phase I and 2 years/$500,000 for phase II. In FY 2000, the NIH made STTR grant awards totaling $21 million, and this was estimated to increase to $24 million for FY 2001.

C. Cooperative Research and Development Agreements

The Federal Technology Transfer Act (FTTA) of 1986 and Executive Order No 12591 mandated the Public Health Service (PHS) to encourage and facilitate collaboration among federal laboratories, state and local governments, universities, and the private sector to assist in the transfer of federal technology to the marketplace. One vehicle for this collaboration is the CRADA between one or more PHS laboratories and one or more nonfederal parties (http://ott.od.nih. gov/NewPages/policy.htm). Under the CRADA auspices, PHS laboratories provide personnel, services, facilities, equipment, or other resources (but not funds) with or without reimbursement to nonfederal parties, and the nonfederal parties provide funds, personnel, services, facilities, equipment, or other resources toward the conduct of specified research or development efforts consistent with the missions of the laboratory. CRADAs confer intellectual property rights on PHS inventions. Although the PHS ensures that outside organizations have fair access to collaborative opportunities, the licensing of federal technologies, and PHS scientific expertise, small businesses located in the United States that agree to manufacture in the United States products developed under the CRADA receive special consideration. However, fair access to CRADAs is not to be considered as synonymous with the term "open competition," as defined for contracts and small purchases.

XIII. CONCLUSIONS

The discovery and evaluation of biomarkers as clinical assessment tools promises many rewards but requires commitment to high-quality research. Recently, increased emphasis has been placed on biomarker research to hasten and improve the assessment of novel pharmaceuticals in clinical trials. Biomarker research is conducted in the watershed of public-funded biomedical science and commercial R&D. Currently, numerous creative approaches to partnering are being undertaken to overcome a variety of clinical research barriers. Biomarker research is not an area of primary focus in either the public or private sectors, yet the knowledge and technologies that emerge from both sectors play important roles in clinical trials. On the other hand, each sector achieves benefits from biomarkers in different capacities, and their utility is better understood as they are introduced into disease management regimens as clinical guideposts for therapy. With this common endpoint in mind, opportunities exist in myriad disease areas for expanded collaborations in biomarker research among public and private research organizations.

Public–private partnerships bring together resources (scientific talent, tools and technologies, specimens, databases, etc.) to expedite the development of effective and safe health technologies—diagnostics, prevention strategies, and therapies. A common feature of the biomarker research is the "precompetitive" areas of discovery and research resource development, which are likely to be the areas where there is the greatest interest and where collaborations will be pursued. Responsible management of research resources, transparent decision making, and effective execution of plans to achieve goals and objectives are keys to successful partnerships in biomarker research.

Exploration of opportunities for partnerships begins with examining current research needs and barriers to clinical therapeutic development. When discovery and resource needs are agreed upon and are mutually beneficial to academic, industry, government, and public nonprofit health research advocacy organizations, public–private partnerships may be an option. Such efforts characteristically encompass, large-scale, complex biomarker discovery and evaluation projects that serve objectives for each of the participating organizations and exceed the capability of being conducted by one segment of the research enterprise.

Some of the anticipated benefits of successful biomarker partnership programs include: an increase in the number of candidate therapies in the development pipeline, an expansion of clinical trials evaluating new therapies for diseases with no or ineffective treatments, improved diagnostic and clinical monitoring capabilities, a decrease in the number of participants needed to evaluate safety and efficacy through more precise clinical measurements, and a shortening of the time frame from preclinical studies to definitive clinical assessment. Based on the examples described here, an argument is made for continued exploration of new research paradigms that exploit discovery opportunities, share resources, and mitigate clinical research barriers to expedite biomarker discovery and evaluation. Although the complexities involved in partnerships between the public and private sectors often seem daunting, a great potential exists for such collaborations in biomarker development, and the numerous creative solutions currently enacted suggest that this potential may be fully realized in the near future. Future analysis will reveal the impact of partnership strategies on pharmaceutical and biotechnology product development and the benefits realized by the public through improved disease detection, prevention, and management technologies.

ACKNOWLEDGMENTS

The author acknowledges the editorial assistance of Robin I. Kawazoe and Charles A. Goldthwaite Jr. for this manuscript.

REFERENCES

1. Biomarkers Definitions Working Group. Biomarkers and surrogate endpoints: preferred definitions and conceptual framework. Clin. Pharmacol. Ther. **2001**, *69*, 89–95.
2. Downing, G.J., Ed.; *Biomarkers and Clinical Endpoints: Advancing Clinical Research and Applications*; Elsevier Scientific: Amsterdam, 2000; 1–7.
3. Rolan, P. Contribution of Pharmocology markers. The contribution of clinical pharmacology surrogates and models to drug development: a critical appraisal. Br. J. Clin. Pharmacol. **1997**, *44*, 219–225.

4. Rolan, P. *Contribution of pharmacology markers. Biomarkers and Surrogate Endpoints: Clinical Research and Applications*; Downing, G.J., Ed.; Elsevier Scientific: Amsterdam, 2000; 27–35.

5. De Gruttola, V.G.; Clax, P.; DeMets, D.L.; Downing, G.J.; Ellenberg, S.S.; Friedman, L.; Gail, M.H.; Prentice, R.; Wittes, J.; Zeger, S.L. Research needs and approaches for the design and analysis of surrogate endpoints in clinical trials: summary of a National Institutes of Health workshop. Control Clin. Trials **2001**; *22*, 485–502.

6. De Gruttola, V.; Wulfsohn, M.; Fischl, M.A.; Tsiatis, A. Modeling the relationship between survival and CD4+ lymphocytes in patients with AIDS and AIDS-related complex. J. Acq. Immune Def. Synd. **1993**, *6*, 359–365.

7. Mellows, J.W.; Kingsley, L.A.; Rinaldo, C.R., Jr.; Todd, J.A.; Hoo, B.S.; Kokka, R.P.; Gupta, P. Quantitation of HIV-1RNA in plasma predicts outcome after seroconversion. Ann. Intern. Med. **1995**, *122*, 573–579.

8. Mildvan, D. Biomarkers as surrogate endpoints in clinical trials for HIV/AIDS: a model for other diseases. In *Biomarkers and Surrogate Endpoints: Clinical Research and Applications*; Downing, G.J., Ed.; Elsevier Scientific: Amsterdam, 2000; 13–17.

9. Kagan, J.M. Biomarkers for HIV/AIDS: challenges ahead. In *Biomarkers and Surrogate Endpoints: Clinical Research and Applications*; Downing, G.J., Ed.; Elsevier Scientific: Amsterdam, 2000; 18–22.

10. Randomised trial of cholesterol lowering in 4444 patients with coronary heart disease: the Scandinavian Simvastatin survival study (4S). Lancet **1994** *344*, 1383–1389.

11. Bush, V. *Science—the Endless Frontier: A Report to the President on a Program for Postwar Scientific Research*; National Science Foundation: Washington, DC, 1980.

12. Smith, R. Vaccines and medicines for the world's poorest: public–private partnerships seem to be essential. Br. Med. J. **2000**, *320*, 952–953.

13. Harrison, P.H. A new model for collaboration: the alliance for microbicide development. Int. J. Gynecol. Obstet. **1999**, *67*, S39–S53.

14. *Creating Global Markets for Neglected Drugs and Vaccines: A Challenge for Public Private Partnership. Global Health Forum I: Consensus Statement*; Institute for Global Health: San Francisco, 2000 (http://www.epibiostat.ucsf.edu/igh/programs/GlobalForum.pdf).

15. Bruntland, G.H. *WHO—the Way Ahead. Statement by the Director-General to the 103rd Session of the Executive Board*; World Health Organization: Geneva, 1999.

16. Buse, K.; Walt, G. Global public–private partnerships: part I—a new development in health? WHO Bull. **2000**, *78* (4), 549–561.

17. Buse, K.; Walt, G. Global public–private partnerships: part II—what are the health issues for global governance? WHO Bull. **2000**, *78* (5), 699–709.

18. Reich, M.R. Public–private partnerships for public health. Nature Med. **2000**, *6*, 617–620.

19. Austin, J. *The Collaboration Challenge: How Non-profits and Businesses Succeed Through Strategic Alliances*; Jossey-Bass: San Francisco, 2000.

20. Pardes, H. The future of medical schools and teaching hospitals in the era of managed care. Acad. Med. **1997**, *72*, 97–102.

21. *The Bayh-Dole Act: A Guide to the Law and Implementing Regulations*; Council of Governmental Regulations, Washington, D.C., 1993.

22. *Patent Policy: Universities' Research Efforts Under Public Law 96-517*; General Accounting Office, Washington, D.C., 1986.

23. *University Research—Controlling Inappropriate Access to Federally Funded Research Results*; General Accounting Office, Washington, D.C., 1992.

24. Bodenheimer, T. Uneasy alliance: clinical investigators and the pharmaceutical industry. N. Engl. J. Med. **2000**, *342*, 1539–1544.

25. Angell, M. Is academic medicine for sale? N. Engl. J. Med. **2000**, *342*, 1516–1518.

26. Stelfox, H.T.; Chua, G.; O'Rourke, K.; Detsky, A.S. Conflict of interest in the debate over calcium-channel antagonists. N. Engl. J. Med. **1998**, *338*, 101–106.

27. Davidson, R.A. Source of funding and outcome of clinical trials. J. Gen. Intern. Med. **1986**, *1*, 155–158.

28. Cho, M.K.; Bero, L.A. The quality of drug studies published in sysmposium proceedings. Ann. Intern. Med. **1996**, *124*, 485–489.

29. Friedberg, M.; Saffran, B.; Stinson, T.J.; Nelson, W.; Bennett, C.L. Evaluation of conflict of interest in economic analyses of new drugs used in oncology. J. Am. Med. Assoc. **1999**, *282*, 1453–1457.

30. DeAngelis, C.D. Conflict of interest and the public trust. J. Am. Med. Assoc. **2000**, *284*, 2237–2238.

31. Stolberg, S.G. Financial ties in biomedicine get a closer look. New York Times, February 20, 2000 p. A1

32. Cho, M.K.; Shohara, R.; Schissel, A.; Rennie, D. Policies on faculty conflicts of interest at US universities. J. Am. Med. Assoc. **2000**, *284*, 2203–2208.

33. Lo, B.; Wolf, L.E.; Berkeley, A. Conflict-of-interest policies for investigators in clinical trials. N. Engl. J. Med. **2000**, *343*, 1646–1649.

34. Moses, H.; Martin, J.B. Academic relationships with industry: a new model for biomedical research. J. Am. Med. Assoc. **2001**, *285*, 933–935.

13

Clinical Operations: Business Principles for Biomarker Applications

John C. Bloom
Lilly Research Laboratories, Indianapolis, Indiana, U.S.A.

I. INTRODUCTION

Providing effective operational support for the application of established and novel biomarkers across all phases of drug development and commercialization is resource intensive, logistically and technically daunting, and, increasingly, a critical success factor for timely regulatory approval and speed to market. Understanding the business principles that underpin successful clinical operations units allows pharmaceutical sponsors to address these challenges and build a competitive advantage. They include standards or attributes of an ideal biomarker database that are required to support claims of efficacy and safety and facilitate successful commercialization. Additional principles include effective sourcing strategies for determining the testing mode (central vs. local), selecting competent service providers, and developing business partnerships that are tailored to the needs of a sponsor's organization and portfolio. Finally, as with all areas of pharmaceutical research and development (R&D), organizational effectiveness is critical to the successful execution of biomarker strategies.

II. BIOMARKER DATABASE STANDARDS

Biomarker data standards are driven by the special needs of pharmaceutical R&D organizations, as regards the use of biomarkers in clinical research. They include the desirable attributes of a biomarker database, which are to be *scientifically defensible, clean (validated), combinable, fully integrated,* and *accessible.* These are the features required to have a database that can be analyzed; reviewed by the sponsor, consultants, and regulators; submitted in a new drug application in a timely fashion; and support demand creation in the marketplace. They each drive specific processes, technologies, and partnerships designed to achieve these goals.

Biomarker data, be they derived from routine or novel laboratory tests, imaging, electrophysiological monitoring, or other diagnostic platforms, are regarded as *scientifically defensible* if the selection of these tests is justified by standards of medical practice, the pharmacology and toxicological potential of the candidate drug, and the pathophysiology of the target disease, among other considerations. Further, the test must be properly validated and performed by accredited laboratories or professional diagnostic services using appropriate standards for quality assurance and quality control. These considerations are discussed in preceding chapters. Pharmaceutical sponsors meet these demands through the use of external consultants and a technical staff of medical specialists and subspecialists. Selecting the right consultant can be challenging, as such a professional must be knowledgeable of both the medical or diagnostic discipline involved and the process of clinical drug development. Sponsors are often challenged by advice that is medically sound from the scientific and patient care perspective, but impractical in the context of a large multicenter clinical trial. Moreover, experts and opinion leaders are not always as attentive to or knowledgeable about the level of documentation and quality assurance required in the highly regulated environment of drug development.

For these reasons, pharmaceutical sponsors often develop relationships with consultants who have both technical and clinical trial expertise required for successful development of a candidate drug. Increasingly, sponsors are recruiting subspecialists in laboratory medicine, radiology, cardiac monitoring, and other key diagnostic disciplines into their organizations, along with paramedical staff that provide both technical and clinical trial process expertise.

Building a biomarker database that is *combinable* and *fully integrated* poses a particular challenge for pharmaceutical sponsors. Two dimensions of "combinability" must be considered: (1) the combination of data from individual sites, regions, or countries in multicenter studies; and (2) the combination of data from multiple studies to provide a longitudinal safety and efficacy database. This requires careful planning and attention to detail to ensure that differences in analytical methods, units, or interpretation do not preclude such data aggregation. Examples of such obstacles include the use of different leads in electrocardiographic (ECG) monitoring, which may preclude a database that can adequately define the effect on QT interval prolongation (see Chapter 4); disparate criteria for tumor regression as measured by sequential computerized tomography scans; coagulation data derived using both percent of control and seconds (these are inherently uncombinable); and hepatic transaminase analyses performed at different temperatures or analytical methods. Efforts to "normalize" such disparate data (percent upper limit of normal, etc.) are sometimes applied, but fraught with problems [1,2].

Sponsors therefore usually attempt to either standardize analytical methods and interpretations or centralize the performance of such assays. The use of standardized methods of ECG collection through business partners that specialize in site management, as relates collection and transmission of ECG data, is an example of the former option, driven by the obvious need for point-of-care assessment. Standardizing laboratory tests that have this requirement, such as those requiring immediate access to the results for patient management, can be particularly challenging. Examples include the assessment of critical care therapeutics or biomarkers employed in chemotherapy units that are used to determine dosage adjustments. More often, however, the challenge of generating laboratory data that are essentially combinable is met through the use of *central laboratories*. Centralized laboratory testing has now become the industry standard, which has spawned a large clinical trial service industry devoted to meeting this need. These and other challenges of providing laboratory support for clinical development are reviewed in Chapter 2.

Standardizing the collection and analysis of biomarker data required to support claims of safety and efficacy across phases I–IV clinical trials is highly desirable for reasons beyond combinability and the integrity of the resultant database. It also facilitates the application of sophisticated information technology (IT) platforms that can capture and manage the large volume of biomarker data generated over the course of the clinical development of a candidate drug. Accordingly, *effective systems*, or IT platforms, are critical to achieving the business objective of building a biomarker database that is clean, combinable, and readily accessible. The ideal system for managing biomarker data has two key attributes: (1) it minimizes study-specific programming; and (2) it provides a high degree of automated cleaning and editing. The value of the latter function cannot be underestimated in a business where time is money. It is estimated that in late-stage development of the average candidate drug, where the net present value and probability of technical success are well defined, each day of delay from registration represents 1–2 million dollars in lost revenue, based on peak sales prior to patent expiration. Progressive, automated validation through biomarker-specific IT tools ensures not only completeness and the opportunity to identify and correct errors in real time, but the ability to minimize the critical period between last patient visit and database lock. The same sophisticated systems that provide these data management functions can also be programmed to provide automated "flags," "alerts," and "panics" that are important for effective patient management, as discussed in Chapter 2. IT platforms for biomarker data also facilitate access to these data, often in real time, by the principal sponsor stakeholders, investigators, and business partners that manage these data on behalf of the sponsor. Increasingly, this is accomplished via the Internet and related data media.

III. CENTRAL VERSUS LOCAL MANAGEMENT OF BIOMARKER SUPPORT

For established biomarkers, such as routine and esoteric laboratory tests, ECG, imaging, and pulmonary function testing, the sponsor is often challenged by the decision as to whether to have these diagnostic tests performed and analyzed *locally* (i.e., by the site using local hospital facilities and professional services) or *centrally*. The latter usually entails management by the sponsor in collaboration with a business partner, as discussed below. Clinical laboratory testing best exemplifies centrally managed diagnostic service support, and is by far the most sophisticated. Factors that should influence the mode of testing (centrally vs. locally managed) are listed in Table 1 and discussed below.

Plans for the data are a key determinant of whether centralized or local testing is indicated. They include whether or not the biomarker data must be managed using IT platforms requiring standardized methods, units, format, and other factors, as well as whether the data will be combined with those that comprise the longitudinal safety and efficacy database. The *timeline for analyses and reporting* is another important consideration, as the aforementioned measures that provide for the automated real-time cleaning and minimize the time to data lock, analyses, and reporting will impact time lines and influence whether investment in centralized management is cost-effective. *Availability of human resources* will also determine whether centralized management of a diagnostic marker using a business partner is needed. The site and database management needs associated with local testing can be resource-intensive, because of the attention required to assure data integrity and fully validate

Table 1 Factors Considered in Selecting Central Versus Local Biomarker Support for Clinical Trials

Plans for data
Time line for analysis and reporting
Availability of human resources
Subcontracting services requiring specimen and data management
Need for data access
Patient management needs
Cost

biomarker and related demographic data captured on case report forms or disparate electronic data platforms.

The *need for subcontracting* services requiring data and specimen management, such as the facilitation of separate bioanalytical (drug and drug metabolite) services, esoteric testing, or third-party peer review of imaging, histopathology, or ECG data, will provide an additional reason to use a centralized diagnostic business partner. The need for *data access* can be an indication or contraindication. Rapid access to clean data requires the standardization and sponsor/diagnostic service provider-interfaced systems discussed previously, and is a strong incentive for centralizing biomarker management. However, the need for "stat" or point-of-care testing that will address patient management needs could preclude the use of this testing mode, or require that local and central testing be done in parallel. Finally, the up-front *cost* of centralized testing is considerably greater than that of local testing. Depending on the above requirements, however, the long-term return on investment of centrally managed diagnostic services relating to quality, speed, and value is often compelling, particularly for multicenter phase II and III clinical trials.

IV. SOURCING STRATEGIES

Another operational challenge in providing routine and special biomarker support in clinical drug development is the selection of service providers and development of business partnerships. Most pharmaceutical sponsors do not regard routine diagnostic support as a core competency. They therefore generally outsource these tasks to vendors that specialize in laboratory, ECG, imaging, and other diagnostic services. Not long ago pharmaceutical sponsors employed the same diagnostic service providers as those used by health care delivery institutions, which were often university or tertiary-care hospitals. The past 10 years has seen a shift to service providers and business partners that specialize in drug development. The emergence of specialized vendors that provide laboratory, cardiac monitoring, imaging, and other diagnostic support for clinical trials was driven by the need for centralized services that are tailored to the aforementioned challenges that pharmaceutical sponsors face in their global clinical drug development efforts.

Developing the business partnerships required to achieve the objectives discussed above poses several challenges. Dimensions of performance that must be assessed (or codeveloped) are listed in Table 2.

Operations of a biomarker business partner must include critical capabilities relating to the service provided. That may entail providing and maintaining ECG instrumentation and data management, specimen transport services, or digitization of radiographic images. Often this requires a global

Table 2   Biomarker Business Partnership Dimensions of Performance

Operations
Technical sophistication
Information technology
Quality assurance and regulatory compliance
Financial considerations (cost)
Other success factors (flexibility, communication, partnership attitude)

presence, or the ability to master the logistics and government regulations required to deliver the service needed in all the major markets. Service gaps in regions such as Japan, Southeast Asia, Eastern Europe, Africa, and South America require the use of multiple service providers or a combination of central and local testing, which defeats the purpose of centralized, standardized testing and limits the return on this substantial investment. Finally, business continuity or stability is critical to the success of biomarker clinical development partnerships that often require 3–10 years of pre- and postregistration clinical research.

Technical sophistication refers to clinical trial process expertise, as discussed previously, as well as the diagnostic subspecialty expertise appropriate to the service provided. Credibility as diagnostic professionals and scientists is essential to ensure both investigator and sponsor confidence. This dimension of performance is particularly critical because of the role these diagnostic services often play in both supporting claims of safety and efficacy of the candidate drug and the management of patients participating in these studies. Technical sophistication may also relate to the analytical instrumentation employed, the ability to develop and validate novel markers or novel applications of established tests, and scale-up capability.

Information technology is becoming an increasingly defining performance dimension of biomarker partnerships, as mentioned above. The automation required today for optimal biomarker data capture, validation (cleaning), event reporting, and other functions requires such features as a predefined database, preaccessioning of specimens, and a highly effective sponsor/vendor interface that provides for electronic data transmission (see Chapter 2). Very sophisticated systems are now required to assist in tasks such as automated site management services; randomization; specimen storage and retrieval; remote data entry; application of sponsor-defined edits, data formatting, and data logic; and other database management support needs. An increasingly important challenge that we as pharmaceutical sponsors face is ensuring that the IT systems our business partners employ are fully validated, in accordance with regulatory requirements such as those of the U.S. Food and Drug Administration on electronic records, signatures, and submissions [3]. Having such validation appropriately documented has become particularly problematic—particularly when collaborating with small, academic-based partners.

Quality assurance and regulatory compliance is another performance dimension that is "nonnegotiable." As discussed previously, this is among the core competencies that distinguish these pharmaceutical diagnostic and biomarker research service providers from those involved principally in health care delivery. The many tasks and parameters that complicate the application of biomarkers in this highly regulated environment include laboratory certification and accreditation; proficiency testing; standard operating procedures; source documentation and audit trail; good clinical/laboratory practices; protocol and data entry, transcription, and transmission; and consistency of methods and processes. The reader is referred to Chapter 10 for further discussion of these topics.

Sourcing strategies of both the sponsor and business partner is another dimension of performance that will influence the success of the partnership. The former has been discussed previously. Included here are sponsor requirements of the service provider, which must be practical and compatible with vendor profitability; how capacity and change management is achieved by the two parties; coordination among global partners, when multiple global partnerships are employed; management and coordination of third-party services by the sponsor and the principal diagnostic service provider; transparency of the sponsor's vendor selection process; and the linkage to overall clinical development sourcing strategies of the sponsor.

The most controversial and challenging of these dimensions is *financial considerations*, or *cost*. This is often a difficult performance dimension to assess, which few pharmaceutical sponsors do well. That is because those assigned to this task often have a marginal understanding of the other performance dimensions, the processes involved, and the critical success factors in providing an effective diagnostic or biomarker research service that supports clinical drug development. This requires an understanding of both up-front and long-term costs and return on investment, such as the cost of generating biomarker data (testing), compared with that of managing, cleaning, analyzing, and reporting the data. Understanding "real value" when assessing complex services that include analytical, medical professional consultation, data and specimen management, site support, and other tasks discussed above can be challenging for procurement functions that often resort to competitive bidding as the key sourcing determinant.

Finally, other factors that influence the success of biomarker business partnerships include the degree of mutual flexibility, communication (particularly in understanding sponsor and partner requirements), a "partnership attitude" that often involves codeveloping successful business practices that are mutually profitable, and the organizational effectiveness of both parties. The latter includes clarity of accountability (responsibility), institutional memory, and standardized processes that preclude having to "reinvent the wheel" for each new project.

V. ORGANIZATIONAL EFFECTIVENESS

Operational effectiveness of biomarker support in clinical development is inevitably influenced by how this function is organized within the pharmaceutical sponsor. Organizational constructs vary substantially across the industry and include biomarker support personnel who are centralized in one division, organized by therapeutic area or product groups and by specific clinical development project teams. Most pharmaceutical sponsors do not have a highly centralized biomarker support function that is responsible for providing these services across all clinical development. An exception to this is in the area of procurement, or sourcing, where a centralized group is often responsible for the contracting of laboratory and other clinical diagnostic services. This is particularly true for central contract diagnostic services such as those supporting clinical laboratory, ECG, and imaging. Some of the same principles and business drivers relating to centralized versus local vendor support discussed previously apply also to biomarker support functions within a sponsor's clinical development organization. For example, organizational constructs involving therapeutic area or team-managed biomarker services generally have greater difficulty in standardizing these processes and building institutional memory as relates best business practices that achieve the aforementioned objectives. The ideal organizational construct is one that is centralized to where standardization relating to protocol development, data collection, management, and reporting; quality standards; and sourcing and business partnership development is achieved while leveraging the technical expertise relating to biomarker research and application that often resides in the therapeutic areas and project teams. Finally, such a centralized biomarker support function must be involved longitudinally throughout the clinical development value chain of protocol design through reporting and submission. This longitudinal *service* function allows the flexibility to modify "up-front" processes in a way that addresses the constantly changing expectations of our regulators and health care providers, as regards the safety and efficacy biomarker data our clinical development units produce. An operational algorithm for a biomarker support function within a clinical development unit used by this author, called the "5 S Model," addresses these organizational challenges. The S's stand for *service, standardization, science, systems,* and *sourcing,* each of which has been discussed above. They are summarized in Table 3.

Table 3 The "5 S" Operational Algorithm for a Clinical Biomarker Support Function

Service
Manage longitudinally across drug development value chain: planning, data lock, submission, commercialization
Customer orientation: project team, investigator site, statistician, regulatory agency
Standardization
Standardize phase-specific common global procedures
Stewardship for biomarker best business practices
Science
Ensure appropriate technical/professional interfaces
Facilitate internal/external consultation
Systems
Maximize available IT platforms
Integrate protocol and data entry system planning
Minimize study-specific programming and data manipulation
Employ virtual platforms that accommodate external data sources and multiple data capture venues
Sourcing
Business case-driven strategy for vendor selection and business partnership development
Integration with key business partners to codevelop required services/processes

VI. CONCLUSION

Sound operating principles for applying clinical biomarkers and managing the resultant data are predicated on the desired attributes of the biomarker database and are critical to the successful registration of candidate drugs. These include having biomarker data that are clean (validated), combinable, accessible, and scientifically defensible in accordance with the standards of medical practice and latest available technology. Testing modes (central vs. local) and strategies for sourcing, business partnership development, and organizational effectiveness are also critical to a sponsor's ability to realize the full potential of these routine and sophisticated experimental tools in demonstrating the safety and efficacy of candidate drugs in a timely and cost-effective fashion, and increasing the probability of regulatory approval and success in the marketplace.

REFERENCES

1.   Cherng, C.N.; et al. Some approaches to the evaluation and display of laboratory safety data. *Proceedings of the biopharmaceutical section*; American Statistical Association: Washington, DC, 1997.

2. Sogliero-Gilbert, G.; et al. A procedure for the simplification and assessment of lab parameters in clinical trials. Drug Inf. J. **1986**, *20*, 279–286.
3. Food and Drug Administration 21 CFR 11: Electronic records; electronic signatures; final rule. Doc 97-6833, 1997.

Index